Series Title: Biological Models

Volume Title: From Knowledge Networks to Biological Models

Editor

Anton Yuryev

Senior Director of Application Science at Ariadne Genomics
Ariadne Genomics
Rockville Maryland
USA

&

Nikolai Daraselia

Chief Scientific Officer at Ariadne Genomics
Ariadne Genomics
Rockville Maryland
USA

CONTENTS

Foreword   *i*

Preface   *iii*

List of Contributors   *vii*

CHAPTERS

1. **Re-Evaluating Androgen Receptor as a Target in Anti-Prostate Cancer Therapy**   **3**

   Dmitry Sivogrivov and Nikolai Daraselia

2. **Role of Ca^{2+}-Mediated Signaling in ALS Pathology**   **24**

   Ekaterina A. Kotelnikova, Mikhail A. Pyatnitskiy, Rachel L. Redler and Nikolay V. Dokholyan

3. **Development of Mechanistic Model for Drug-Induced Cholestasis and its Applications for Drug Development**   **73**

   Nikolai Daraselia, Pat Morgan and Anton Yuryev

4. **Pathways Disturbed in Duchenne Muscular Dystrophy**   **104**

   Maria A. Shkrob, Mikhail A. Pyatnitskiy, Pavel K. Golovatenko-Abramov and Ekaterina A. Kotelnikova

5. **Mechanism of Synergistic Carcinogenesis from Hypergastrinemia and Helicobacter Infection**   **131**

   Anton Yuryev

6. **Sub-Network Enrichment and Cluster Analysis Reveal Possible Pathways for Cetuximab Sensitivity**   **151**

   Mikhail A. Pyatnitskiy, Maria A. Shkrob, Nikolai D. Daraselia and Ekaterina A. Kotelnikova

   Index   **173**

FOREWORD

Molecular Biology rapidly evolves from experimental science to computational discipline. This transformation is fueled by simultaneous advances in modern computing and explosion of global molecular profiling methods. Typical molecular profile contains tens of thousands data points and its interpretation relies on the relational database storing formalized knowledge about molecular interactions. The development of computerized knowledgebase for pathway and network analysis started in the beginning of this century in response to the advances in DNA hybridization microarray technology which allowed simultaneous mRNA expression measurement for all genes in a biological sample. In 2003 Ariadne Genomics pioneered MedScan information extraction technology in order to find statements about molecular interactions in scientific literature and to automatically populate the knowledgebase with extracted information. MedScan is highly accurate technology which reliably converts the enormous amount of literature accumulated during more than 60 years of research into the knowledgebase suitable for computational analysis. MedScan makes Pathway Studio a unique software product which provides tools for navigating the most comprehensive knowledgebase in molecular biology.

Scientific literature proved to be extremely rich source of molecular interaction data which suffers from large number of errors and omissions. Biological data is intrinsically ambiguous not only due to the technical noise from the experimental set up but also due to the natural genetic variability and genetic linkage in biological samples. Genetic variability makes a response to the same environmental changes unique in every biological sample. Genetic linkage causes every response to include non-specific components which are functionally irrelevant to the response. High level of noise in the knowledgebase is exuberated with the noise in high-throughput molecular profiling data that is analyzed using the knowledgebase. Hence, the necessity to sift through the knowledgebase lead to the development of statistical algorithms capable of finding key regulatory events relevant to biological response or cell process in focus. I have worked at Ariadne Genomics on developing sub-network enrichment analysis which has become the major tool for interpreting raw molecular profiling data. I am happy to see how extensively SNEA is used throughout this book for making inferences from gene expression microarray data providing the foundation for building mechanistic models.

Building predictive mechanistic models in biology requires multiple expert skills including thorough understanding of context and experimental approaches used for measuring interactions in the knowledgebase, thorough understanding of the limitations of high-throughput molecular profiling technologies, expert understanding of cellular processes involved in disease or biological response, and thorough understanding of statistical algorithms enabling the knowledge inference. Very few people in the world possess this combination of skills. Therefore I am not surprised that the book is written almost entirely by Ariadne team who also took advantage of the powerful graphical interface for pathway visualization and construction available in Pathway Studio. This book provides readers with the deep insights into how the raw biological data can be converted into predictive *in silico* models.

Andrey Sivachenko, Ph.D.

Broad Institute
Boston, MA
USA

PREFACE

Chapters in this book describe building mechanistic models for various human diseases and conditions. While each chapter provides novel insights into the disease mechanism and should be of interest to any expert in this disease, we note that the authors in this book have never published articles about the disease described in their chapter and have never performed any experiments to study the disease. All authors have learned and advanced the understanding of the disease mechanism either through analysis of knowledge networks or through analysis of publicly available gene expression datasets using knowledge networks. All chapters also have in common the use of Pathway Studio software from Ariadne Genomics. Pathway Studio provides access to the biological knowledge networks and tools for their navigation and analysis. Most knowledge in the Pathway Studio database is extracted automatically from scientific literature using MedScan information extraction technology. While MedScan is thoroughly described in publications from Ariadne Genomics, the goal of this book is to show how to use the extracted information for knowledge inference, for building mechanistic models, and for learning how to use the model and knowledge networks to make more informed predictions about disease targets and biomarkers. We emphasize that while every model in this book required MedScan-extracted knowledge networks, Pathway Studio also allows import and navigation of additional knowledge from other sources and databases. Some examples of additional knowledge - protein homology network or network of physical interaction imported from public PINA database - are described in the chapters about cholestasis and gastric cancer models.

So what are "knowledge networks"? There are a couple of ways to answer this question. The analogy with the computer science term "Semantic Web" is the first that come to mind. For readers with a biological background, another definition of "molecular biological knowledge networks" can be *compressed, formalized representation of the knowledge about biological molecular interactions described in scientific literature.* Statements about molecular interactions, molecular function, and about molecule roles in disease and other phenotypes are scattered among millions of articles published by the scientific community in the last 60 years. MedScan converts such statements into semantic triplets, *e.g.*, "A regulates B" or "C binds B", in order that they can be imported into a relational database. The Pathway Studio database generated by MedScan 5.0 technology contains more than 2.5 million unique relationships described in more than 18 million molecular biological articles. Knowledge networks stored in the Pathway Studio relational database provide instantaneous access to the knowledge generated by entire molecular biological research that has been supported by trillions of dollars of investment.

The compression of quintessential molecular biological knowledge into semantic triplets allows both a quick overview for users and rapid traversing using network navigation algorithms. By bringing together in one database information extracted from disparate knowledge domains, Pathway Studio enables individual domain experts to make analytical connections that have been previously unnoticed. It allows the making of statistically sounder conclusions that are based on all published observations rather than on the limited set of papers familiar to only one expert. There are three major domains in biomedical knowledge: physical and regulatory molecular interactions measured in basic academic molecular biological research; pharmacological effects and drug interactions published by medicinal chemists from the pharmaceutical industry and pharmacology and translational medicine departments in academia; disease - related molecular changes published by clinicians and medical doctors. Medical doctors rarely know molecular biology and basic scientists usually do not know much about pharmaceutical research. Bringing together molecular interactions and clinical observations are essential, however, for building a molecular mechanism of a disease. Knowledge about drug mechanisms is necessary for finding new drugs based on the mechanistic disease models.

Any given drug or disease may affect the activity of dozens and often hundreds of biological molecules. While contemporary high-throughput molecular profiling technologies, such as gene expression microarrays, can measure global molecular response, the interpretation of observed profile requires an overview of thousands of publications describing individual interactions between genes and proteins in the profile.

Such intermolecular dependencies are often measured in individual academic labs independently from clinical or drug research. Another example of separation in biomedical knowledge is the context specificity of observed molecular interactions and functions. Due to the high cost of molecular biological experiments, individual molecular interactions are usually measured only in the context of one tissue, organism or condition. Most of these context-specific interactions can be used for building a model for another disease or to explain the molecular profile measured in a different tissue or organism. While borrowing interactions from another organism or tissue is a common practice for building biological models, the search for such interactions through biomedical literature would be a daunting task without Pathway Studio and its knowledge networks.

This book is written by Pathway Studio experts to show how one can leverage the information integrated into the knowledge networks for building mechanistic models. While the knowledge networks consititute a global compendium of molecular interactions observed by entire molecular biological research, the mechanistic model of a disease, phenotype, or trait contains only a subset of such interactions. This subset must be sufficient to explain all or a majority of molecular observations about the condition. The first step in building a model is collecting all observations from various scientific publications and enriching it with the results obtained by a global molecular profiling experiment. For many complex diseases, such as cancer, this effort leads to the collection of several thousand proteins affected by the disease state. The process of model building can be described as complexity reduction of the observed molecular profile for a given disease or condition. You will learn from the book chapters that even changes in thousands of genes and metabolites affected by disease can be explained by the activity change in only a few biological pathways.

Three chapters in the book use public gene expression datasets profiling the disease state and comparing it to healthy control "normal" state. The principal technique of reducing complexity of a molecular profile is called sub-network enrichment analysis (SNEA). In the case of gene expression, SNEA uses the expression regulatory knowledge network to find transcription factors and other regulators responsible for the biggest changes observed in the experiment. You will see that, throughout the book, SNEA regulators can often be mapped onto one or several canonical pathways, indicating that pathway changes its activity in the disease state. Due to the small number of pathways known for the human organism, it is not always possible to map significant expression regulators identified by SNEA. Therefore, the last two chapters suggest other techniques - regulator clustering and pathway reconstruction - to classify expression regulators into a smaller number of functional communities in order to further reduce the complexity of the molecular profile.

We hope that the examples from this book will allow readers to start building models for their disease or phenotype of interest. The book starts with simpler chapters that use knowledge networks to review the state-of-the-art in a disease field. The last chapters describe more complicated applications of knowledge networks for building disease models by analyzing public gene expression datasets. Some chapters go beyond model building. Once the disease model is built, it can be used for more accurate prediction of biomarkers, repositioning of existing drugs, target selection for future drugs, and design of personalized therapy using the same knowledge networks available in Pathway Studio.

Anton Yuryev

Senior Director of Application Science at Ariadne Genomics
Ariadne Genomics
Rockville Maryland
USA

&

Nikolai Daraselia

Chief Scientific Officer at Ariadne Genomics
Ariadne Genomics
Rockville Maryland
USA

List of Contributors

Dmitry Sivogrivov

Ariadne Genomics
E-mail: siv-dmitrey@ariadnegenomics.com

Ekaterina Kotelnikova

Ariadne Genomics
E-mail: ekotelnikova@ariadnegenomics.com

Nikolai Daraselia

Ariadne Genomics
E-mail: nikolai@ariadnegenomics.com

Maria Shkrob

Ariadne Genomics
E-mail: shkrob@gmail.com

Anton Yuryev

Ariadnc Genomics
E-mail: anton@ariadnegenomics.com

Mikhail Pyatnitskiy

Ariadne Genomics

CHAPTER 1

Re-Evaluating Androgen Receptor as a Target in Anti-Prostate Cancer Therapy

Dmitry Sivogrivov* and Nikolai Daraselia

Ariadne Genomics Inc., Rockville, MD, USA

Abstract: Androgen deprivation therapy (ADT) is the gold standard for advanced stages of prostate cancer due to the good patient responding to this treatment. Nevertheless, in most cases the disease acquires hormone-refractory status which is considered incurable. The aim of this study is to draw a mechanistic model of androgen receptor (AR) signaling in normal prostate and cancerous tissue in order to better understand the causes of treatment failure. With assistance of Pathway Studio software (Ariadne Genomics Inc., USA) we have created overview pathway showing principal signaling cascades disturbed during prostate cancer development and progression. We further discuss the role of androgen receptor in homeostasis of prostate normal cells and in deregulating signaling of cancer cells. Due to AR involvement not only in proliferation, but also in apoptosis, invasive growth and in cell-to-cell communication, we hypothesized that ADT in some cases may contribute to the development of the castration resistant cancer. We review that upon androgen deprived conditions impaired AR activity can lead to the disturbances in multiple pathways thereby influencing global homeostasis of the prostate. We concluded that potential of AR to potentiate prostate cancer rather than inhibiting it should be taken into consideration when choosing ADT as an option for prostate cancer management.

Keywords: Androgen receptor, prostate cancer, androgen deprivation therapy, Pathway Studio software, AR signaling, AR target, prostate cancer overview pathway, prostate specific antigen, castration resistant prostate cancer, tumor suppressor, oncogene, cell proliferation, apoptosis, cancer metastasis, testosterone, AR-dependent cell, cell homeostasis, prostate cancer treatment, cell cycle, AR antagonist.

INTRODUCTION

Prostate cancer is the most common cancer in men. It is the second leading cause of death among all cancer types in men in Europe and North America [1] and the

***Address correspondence to Dmitry Sivogrivov:** Ariadne Genomics Inc., 9430 Key West Ave., Rockville, MD 20850, USA; E-mail: dsivogrivov@itcwin.com

Anton Yuryev and Nikolai Daraselia (Eds)

sixth leading cause worldwide [2]. Interestingly, men from some regions (*e.g.*, Asia) have considerably lower risk of prostate cancer development compared to Western countries (fewer than 10 and about 100 incidences per 100, 000 men, respectively) [2]. Other risk factors are older age (over 60) and family history (hereditary causes) [3]. Prostate cancer is considered to be a slow growing malignancy, and sometimes patients die from another disease without the appearance of prostate cancer symptoms [4]. Some cases, however, are quite aggressive and produce metastases, more often to lymph nodes [5] and bones [6]. The main treatments of primary tumors are surgery (radical prostatectomy) and radiation therapy [7, 8]. However, in approximately 35% of the cases the cancer recurs [9] and in this case the most used treatment is hormonal therapy (ADT—androgen deprivation therapy) aimed at decreasing testosterone level and, as a consequence, androgen receptor activity [10]. Approximately 80% of patients respond to this treatment but the median duration of the response is only 12-18 months and then an advanced stage of the disease is developed that is known as hormone-refractory or castration-resistant prostate cancer [11]. This stage is considered to be incurable with a median survival rate for hormone-refractory prostate cancer of two years [12]. Understanding of the biochemical background of advanced stages of prostate cancer will enable us to manage this malignancy more effectively.

STAGES OF THE DISEASE AND CURRENT TREATMENT

Proper staging of prostate cancer plays an important role in therapy design [13]. There are three stages of the disease: primary tumor (localized prostate cancer), locally advanced prostate cancer, and metastatic prostate cancer. A more detailed characterization of the tumor is called the TNM staging system (abbreviated from Tumor, Nodes, and Metastases) [14]. According to this method, T1 and T2 stages correspond to localized prostate cancer, whereas T3 and T4 correspond to metastatic prostate cancer. Each stage in the TNM system is divided into subgroups that more accurately describe tumor size and position [14]. Histopathological evaluation of the tumor is usually carried out according to the Gleason Grading System when a sample of the prostate tissue (biopsy) is available [15, 16]. In this technique the evaluating parameter is the histologic grade that characterizes the degree of changes in the prostate tissue. A grade from 1 to 5 is assigned to two most common tumor patterns and the sum of two grades is an overall Gleason Score (ranging from 2 to 10). TNM stages and the Gleason Score are often used together to better describe the level of tumor development [17]. Prostate-specific antigen (PSA) testing used in conjunction with the previously described tumor characterizations is helpful for patient stratification into risk groups - for example, in assessing the risk of biochemical recurrence according to the D'Amico classification [18], which has three risk groups for recurrence: low, intermediate, and high. This stratification system is also helpful in making a choice for optimal treatment: patients with low risk sometimes require no treatment at all, whereas patients with high risk require more aggressive treatment [8]. Despite the existence of the patient stratification system, there is no good treatment for all men diagnosed with prostate cancer. The treatment option depends on the patient's age and health as well as the treatment's potential side effects [8].

Current treatment of localized prostate cancer may comprise surgery, radiation therapy or active surveillance/watchful waiting [19]. Radical prostatectomy involving removal of prostate gland and some adjacent tissue is prescribed for patients with good health. For patients with localized or locally advanced prostate cancer, radiation therapy is an option [20]. Several types of radiotherapy are used for prostate cancer treatment: external beam radiation therapy, proton therapy, and brachytherapy [21]. Prostatectomy and radiotherapy are relatively effective treatments with a good overall survival rate. However, they have some side effects such as urinary incontinence, erectile dysfunction, acute cystitis, proctitis [8, 22]. Because of these complications or due to patient's specific contraindications active surveillance/watchful waiting is becoming a popular option for men with prostate cancer [8]. This approach does not implicate active intervention and is based on continuous monitoring (*e.g.*, digital rectal examination, PSA testing, and/or biopsy) for signs of progression. Its advantage is the avoidance of unnecessary treatment that sometimes results from overdiagnosis [23]. This option is recommended in cases of tumors with low grade, low risk, and low volume or in case of limited life expectancy of older patients with high grade or metastatic disease when the treatment is considered to be ineffective [8, 24].

For locally advanced and metastatic prostate cancer, androgen deprivation therapy (ADT) is the gold standard of the treatment [10, 25, 26]. ADT is performed by surgical (orchiectomy) or medical (pharmacologic) castration to reduce the availability of circulating testosterone. While surgical castration is irreversible and is often avoided for this reason, hormonal therapy can be stopped at any time, leading to a rise in testosterone level [26]. Pharmacologic castration can be achieved either by blockade of testosterone biosynthesis (GnRH agonists and antagonists) or by reducing the activity of the androgen receptor by AR antagonists (bicalutamide, flutamide, nilutamide) [10, 25]. These approaches are used sometimes in combination to achieve a better effect [24]. Although ADT is effectively used to stop the tumor progression, its main disadvantage is the development of the androgen-independent state that is currently considered incurable [27]. Furthermore, there is evidence that androgen deprivation therapy has numerous adverse effects such as sexual dysfunction, hot flashes, weight gain, and osteoporosis. It also can increase the risk of cardiovascular diseases and diabetes [25]. These disadvantages, along with the fact that about 70% of hormone therapy is prescribed to patients without proven benefit [28], necessitate the development of novel therapies for advanced stages of prostate cancer.

We conclude that the current state of the art allows a cure for the majority of localized prostate cancer cases providing good overall prognosis after treatment, while advanced stages of the disease, especially metastatic cases, are generally incurable and require novel treatment options.

BIOCHEMICAL CHARACTERIZATION OF PROSTATE CANCER - OVERVIEW PATHWAY

Prostate cancer is characterized by considerable heterogeneity, especially during the progression to hormone-refractory status [29, 30, 31]. To better understand the biochemistry of prostate cancer cell(s), we have summarized the main signaling cascades affected by the disease in the form of the pathway (Fig. **1**). The overview pathway depicts signaling cascades that can be affected in cancer cells but are not necessarily affected in every case. The description of some signaling cascades that play a role in prostate cancer follows.

PI3K/Akt: One of the deregulated signaling cascades in prostate cancer cells is PI3K/Akt pathway that is found to be highly activated in advanced cases of prostate cancer. The main cause of constitutive activation of Akt is the loss of PTEN function in advanced cancer because negative regulation of the PI3K pathway is primarily accomplished through the action of tumor suppressor PTEN [30]. Akt kinase regulates cell survival and cell cycle. For example, Akt negatively regulates apoptosis by inhibiting procaspase-9 and BAD and positively regulates cell cycle by inhibiting p21 and p27, thus enabling cell cycle progression [30]. Furthermore, Akt kinase can also regulate androgen receptor activity. In the normal prostate AR apparently balances proliferative and proapoptotic programs, while in cancerous tissue Akt protein can phosphorylate and inactivate the androgen receptor [32], thereby permanently switching on a pro-survival program [33, 34]. Another target of Akt kinase is the enhancer of zeste homolog 2 (EZH2) that is a transcriptional repressor whose expression in prostate cancer correlates with progression to hormone-refractory and metastatic disease [35].

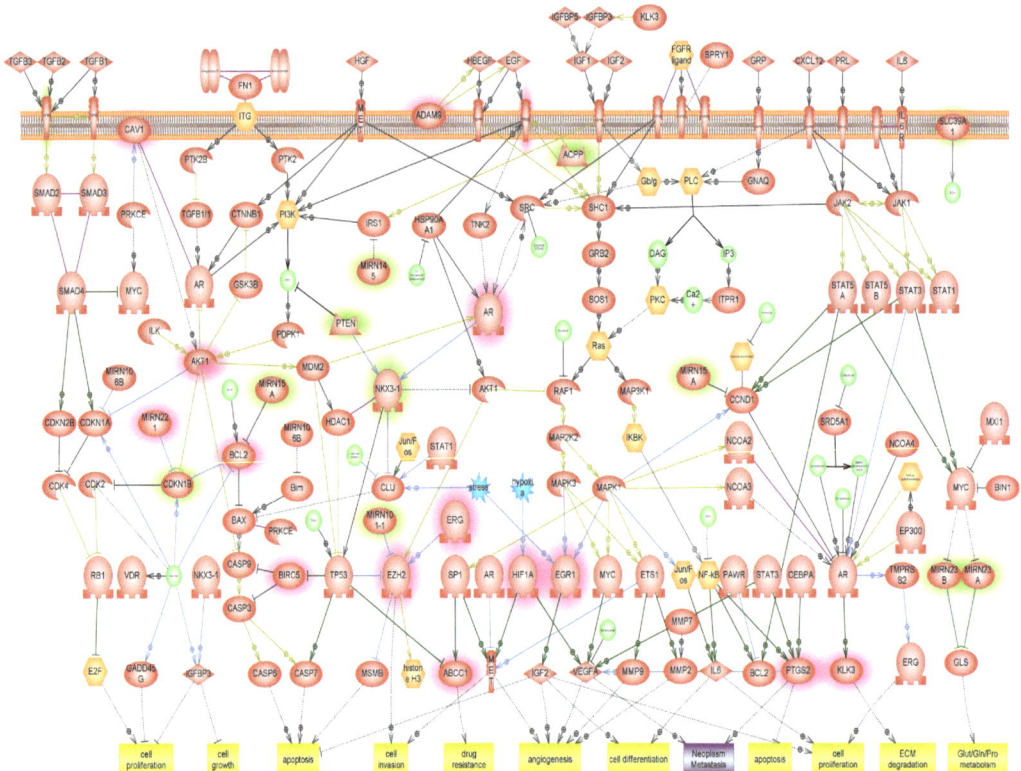

Figure 1: A prostate cancer overview pathway created in Pathway Studio software (Ariadne Genomics, Inc., USA). Entities overexpressed in prostate cancer are in fuchsia color and entities with a decreased or no expression are in green. Relationships between entities are confirmed by manual review of the references available in the program interface.

MAPK: Another signaling cascade that is deregulated in prostate cancer is the MAP kinase (MAPK) pathway. This pathway controls such processes as proliferation, angiogenesis, metastasis, and resistance to anticancer treatments. Several studies have shown that MAPK activity correlates to the progression of an advanced and hormone-independent prostate cancer [36]. The main probable cause for permanent activation of the MAPK cascade in the prostate is increased levels of autocrine and paracrine growth factors. Additionally, prostate cancer cells are characterized by overexpression of some growth factor receptors (*e.g.*, ERBB2), that also activates growth factor pathways [37]. The diverse array of growth factors, cytokines, and proto-oncogene products such as epidermal growth factor (EGF), insulin-like growth factor-1 (IGF-1), and interleukin-6 (IL-6) deliver their growth-eliciting stimuli through the activation of small G-protein Ras and subsequently of the activation of the MAPK cascade [36]. Phosphorylation of ERK1/2 leads to activation of various transcription factors such as Myc, ETS-1, EGR-1, SP-1, and Jun/Fos and the modulation of downstream cell processes [30].

IL-6: One of the transcriptional targets of MAPK signaling is interleukin-6, which can play a role in the progression of prostate cancer through several signaling cascades (JAK-STAT, PI3K/Akt and MAPK) [38]. Constitutive STAT3 activation can contribute to cancer progression by enhancing AR transcriptional activity when the concentration of circulating androgens is low due to ADT therapy. STAT3 has been shown to form a complex with the androgen receptor and modulate its activity [39]. IL-6 can also enhance AR activity through the activation of the MAPK cascade, which can directly phosphorylate AR and its coregulators [36].

TGF-beta: In a normal prostate, TGF-beta is predominantly produced by prostatic stromal cells and functions as a paracrine inhibitor of normal prostate epithelial cell proliferation. It is thought to be a mediator of castration-induced epithelial apoptosis [38]. TGF-beta signaling occurs through its receptor and SMAD proteins leading to transcriptional activation of cell cycle inhibitors p15 and p21 [40] and inhibition of proto-oncogene Myc [41]. In prostate cancer cells, the decreased sensitivity to TGF-beta is due to reduced levels of the TGF-beta receptor, resulting in a proliferative effect.

NED: Another hallmark of prostate cancer is neuroendocrine differentiation (NED), which leads to the formation of neuroendocrine-like cancer cells [42, 43]. Neuroendocrine tumor cells, unlike non-neuroendocrine tumor cells, do not express AR and are likely androgen-independent. Therefore, they may survive and continue to function in the androgen-deprived environment and establish autocrine and paracrine networks stimulating androgen-independent growth of prostate cancer [43]. For example, prostate cancer cells often express the gastrin releasing peptide (GRP, bombesin) receptor on their surfaces [44] through which GRP (as a product of NE cells) exerts its mitogenic effect.

microRNAs: There are also a number of microRNAs that play a major role in the development, invasion, metastasis and prognosis of prostate cancer. Several miRNAs and their targets have been discovered to express abnormally in prostate cancer, including some that are oncogenic (miR-106b, miR-221) and tumor suppressive (miR-15a, miR-145, miR-101-1) [45, 46].

Prostate cancer progression to the hormone-refractory and metastatic stage due to androgen deprivation therapy makes it difficult to manage this type of cancer through existing approaches. Deregulation of multiple signaling cascades confirms this fact. There is a need to find more relevant therapy from the very beginning of the treatment.

THE ROLE OF *AR* IN NORMAL PROSTATE AND CANCEROUS TISSUE

The androgen receptor plays an important role both in the normal prostate gland and in prostate cancerous tissue. The latter observation suggests fighting the disease by silencing androgen receptor signaling during androgen deprivation therapy [47]. Indeed, this approach slows down the growth of cancer, but cancer inevitably acquires an androgen-independent status and stops responding to hormonal therapy [10]. This indicates that the AR transcriptional program is very important in prostate tissue homeostasis, and therefore the disruption of this program can lead to the transformation of many signaling cascades. The goal of this chapter is to better understand processes affected by androgen receptor signaling in the prostate cells.

The androgen receptor is absolutely required for the development and functioning of the prostate gland. This is confirmed by the absence of prostate in mice with knockout mutations of AR [48] and in men with mutations that completely abolish AR function [49]. The androgen receptor is primarily expressed in secretory epithelial cells and to a lesser extent in stroma, whereas basal epithelial and neuroendocrine cells are AR negative [38]. Androgen receptor activity is dependent on steroid hormones, particularly upon testosterone and the more potent 5α-dihydrotestosterone (DHT) serving as ligands for the receptor. Upon a castration condition (in the absence of androgens) most of AR dependent prostate cells undergo apoptosis, resulting in a loss of 70% of secretory epithelial cells [50]. Activated AR turns on the transcriptional program which appears to hold the balance between proliferative and apoptotic processes. What follows are examples of AR regulation of these and other cell processes.

One well-defined target of the androgen receptor is prostate-specific antigen (PSA). Transcription of PSA (also known as kallikrein, KLK3) is initiated when the ligand-activated androgen receptor binds to a region in the PSA promoter that contains an androgen-responsive element (ARE) (Fig. **2**).

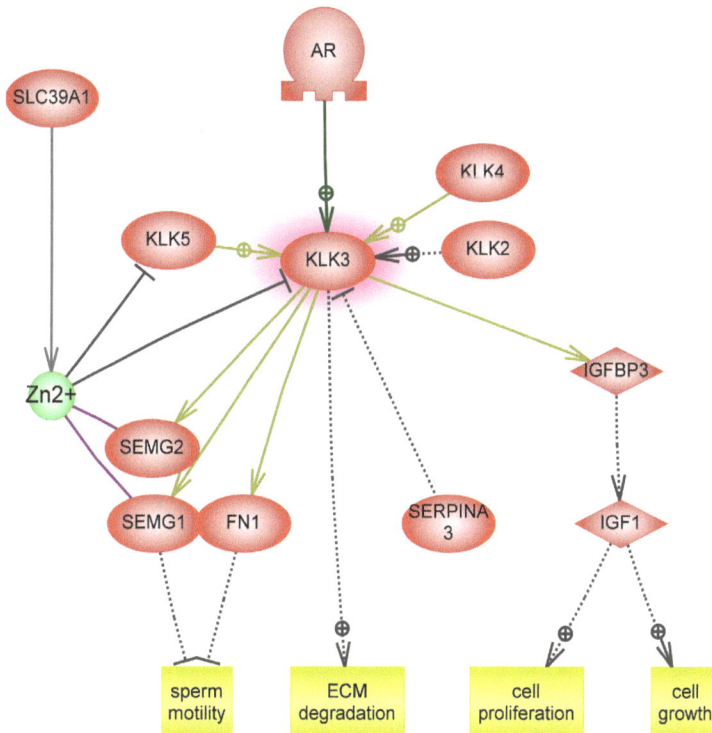

Figure 2: AR signaling through PSA (KLK3).

In the normal prostate, kallikreins are activated in secretory epithelial cells but are inhibited by Zn^{2+} in a reversible manner. Having once appeared in ejaculate, they are reactivated due to Zn^{2+} redistribution to semenogelins, in order to liquefy the seminal clot by proteolysis of semenogelins and fibronectin. These processes lead to the release of motile spermatozoa [51]. Such a program is possible due to the prostate's ability to accumulate cellular levels of zinc that are three- to tenfold higher than other mammalian cells *via* the working of ZIP1 and ZIP2 proteins [52]. During prostate cancer initiation and progression when zinc-accumulating apparatus is often downregulated, the decrease of Zn^{2+} levels may result in an increase in KLK3 activity and increased degradation of extracellular matrix components and insulin-like growth factor-binding proteins. These events could enhance prostate cancer progression and metastatic potential [53].

Besides playing a role in the physiology of seminal fluid, the androgen receptor regulates proliferative and apoptotic processes in AR-dependent cells. AR-regulated signaling cascades, whose influence these processes are depicted in Fig. **3**.

One of the targets of the androgen receptor is *NKX3-1* gene (Fig. **3a**). The homeodomain-containing transcription factor NKX3-1 is a putative prostate tumor suppressor that is expressed in a largely prostate-specific and androgen-regulated manner [54]. Loss of NKX3-1 protein expression is a common finding in human prostate carcinomas and prostatic intraepithelial neoplasia. One of the known roles of the NKX3-1 tumor suppressor is to modulate activity of the p53 protein *via* interaction with HDAC1 and removing it from the p53-MDM2-HDAC1 complex [55]. As a result, p53 is acetylated and activated to induce apoptosis. Another possible way in which androgen receptor could regulate cell proliferation and growth is through increasing expression (*via* NKX3-1) of an insulin-like growth factor binding proteins that are responsible for the decrease in bioavailability of the insulin-like growth factor (IGF-1) [56]. Furthermore, NKX3-1 possibly serves as a transcriptional repressor which decreases the expression of the pro-oxidant enzyme quiescin Q6 [57]. The loss of NKX3-1 function due to mutation can lead to an increased level of quiescin Q6 and tumor initiation through increased ROS accumulation. The NKX3-1 itself can be regulated by phosphorylation, and casein kinase II participates in this process a)

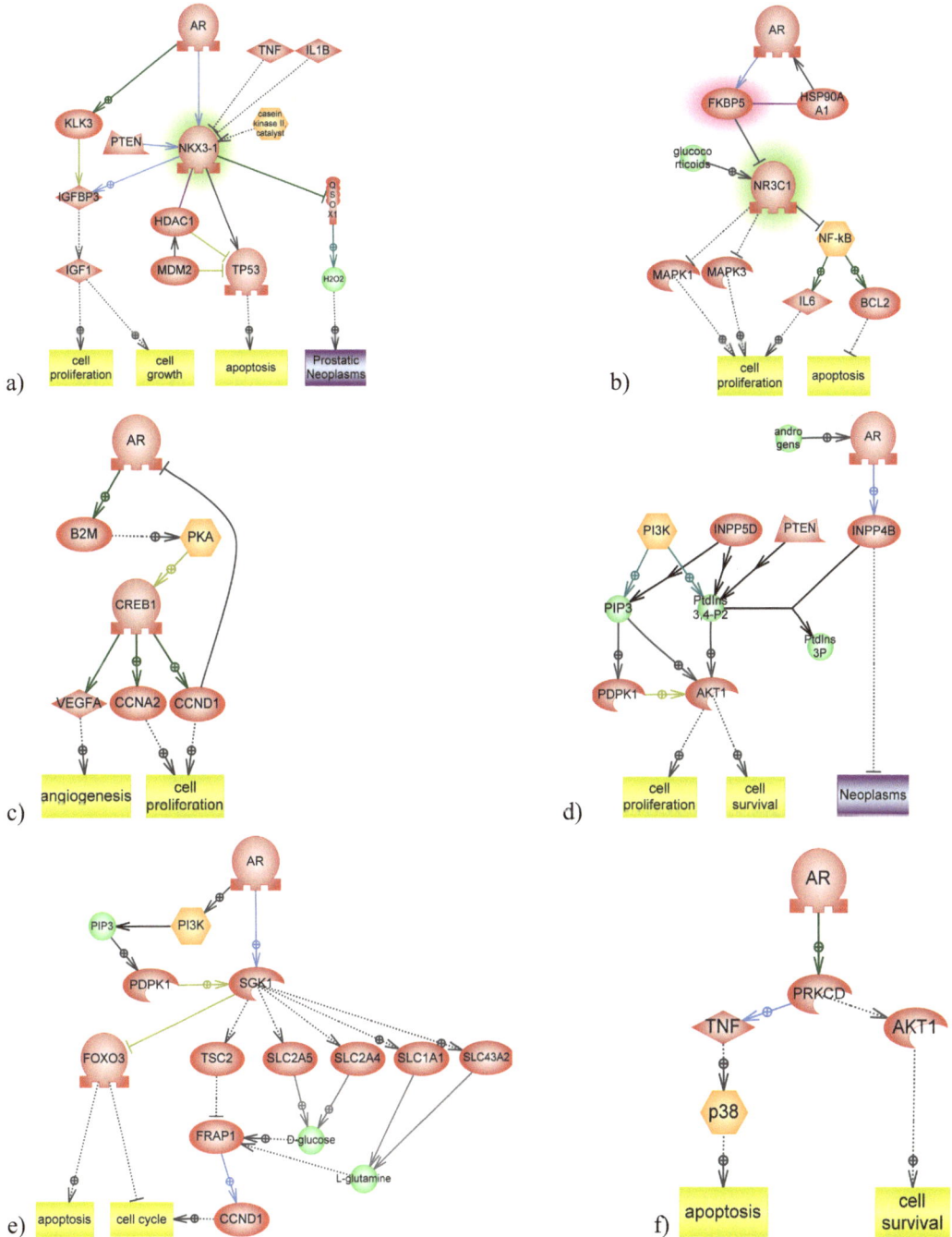

Figure 3: The role of the androgen receptor in the regulation of proliferation and apoptosis. AR signaling through NKX3-1 (a), FK506 binding protein 5 (b), beta-2-microglobulin (c), inositol polyphosphate-4-phosphatase (d), serum/glucocorticoid regulated kinase 1 (e), and protein kinase C delta (f).

enhancing stability of NKX3-1 and also changing its DNA binding affinity [58]. There is also evidence that chronic inflammation can induce cancer initiation and progression because inflammatory cytokines such as TNF-alpha and IL-1beta can accelerate NKX3-1 protein loss [59].

Another target of the androgen receptor is the *FKBP51* gene (Fig. **3b**). FK506-binding immunophilin 51 (FKBP51) is a member of the immunophilin protein family, which plays a role in immunoregulation and basic cellular processes involving protein folding and trafficking [60]. FKBP51 associates with heat shock protein 90 and appears in functionally mature steroid receptor complexes [61]. This could enhance the stability of the androgen receptor. Another possible role of FKBP51 in AR signaling is regulation of glucocorticoid action. Under physiological conditions FKBP51 protein participates in the formation of the mature glucocorticoid receptor (GR) complex [62]. However, overexpression of FKBP51 was found to reduce GR hormone binding affinity. This could lead to increased cell proliferation and inhibition of apoptosis because GR can regulate these processes through the mechanism of transrepression. In particular, GR can complex with NF-κB or AP-1 transcription factors preventing them from regulating target genes [63]. GR can also inhibit the MAPK cascade. There is evidence that GR functions as a tumor suppressor in the prostate and is often downregulated or absent in prostate cancer specimens and cell lines [64].

The androgen receptor can also signal through the regulation of transcription of beta-2-microglobulin protein (Fig. **3c**). Beta-2-microglobulin (B2M) is a serum protein found in association with the major histocompatibility complex (MHC) class I heavy chain on the surface of nearly all nucleated cells. B2M is also known to be a mitogen and is capable of increasing the growth of prostate cancer cells and regulating the expression of hormone/growth factor receptors [65]. The gene *B2M* has a putative androgen response element and is known to be overexpressed upon androgen stimulation [66]. It is proposed that B2M can activate cAMP-dependent protein kinase A possibly through binding to the G protein-coupled receptor [67]. Subsequent phosphorylation and activation of transcription factor CREB lead to expression of cyclin A, cyclin D and VEGF and increase cell proliferation, survival and angiogenesis.

There is also crosstalk between the AR and PI3K/Akt pathways (Fig. **3d**). It is generally accepted that phosphatidylinositol (3,4,5)-triphosphate (PIP3) binds to and activates Akt kinase, whereas tumor suppressor PTEN phosphatase dephosphorylates PIP3 (to phosphatidylinositol (4,5)-diphosphate, PIP2), thus decreasing activity of the Akt pathway [68]. Akt can also be activated by phosphatidylinositol (3,4)-bisphosphate (PtdIns 3,4-P2) that in turn is dephosphorylated by PTEN [69]. Another phosphatase dephosphorylating PtdIns 3,4-P2 is INPP4B (inositol polyphosphate-4-phosphatase), which is considered to be a tumor suppressor in epithelial carcinomas [70]. It was shown that the androgen receptor induces the expression of the INPP4B in prostate cancer cells, and INPP4B depletion activates Akt and increases cellular proliferation [71]. This can serve as evidence of adverse effects accompanying androgen deprivation therapy when the negative regulation of AR leads to prostate cancer transformation into hormone-refractory status.

The androgen receptor can also modulate the activity of serum and glucocorticoid-regulated kinase 1 (SGK1), thus regulating cell cycle progression and apoptosis (Fig. **3e**). SGK1 belongs to the AGC family of serine/threonine protein kinases, which also includes protein kinase C and Akt [72]. The SGK1 is an androgen-regulated target gene, and its mRNA level is upregulated after androgen stimulation [73]. SGK1 can impact cell physiology through various mechanisms. On the one hand, it can regulate the cell cycle and apoptosis indirectly through inhibition of the FOXO3 transcription factor that is responsible for cell cycle arrest and apoptosis [74]. On the other hand, SGK1 can control cyclin D and entry into the cell cycle by regulating the activity of the mammalian target of rapamycin (mTOR) protein [75]. This regulation can be performed by activating various cellular transporters and the subsequent influx of nutrients and amino acids which are positive regulators of mTOR [76].

Regulation of PKC-delta by the androgen receptor is depicted in Fig. **3f**. Protein kinase C delta (PKC-delta) is a member of a family of serine- and threonine-specific protein kinases and is involved in regulation of growth, apoptosis, and differentiation of a variety of cell types [77]. It was shown that the PKC-delta gene promoter has an androgen response element and is highly responsive to androgens [78]. Activated PKC-delta can accomplish a pro-apoptotic program *via* stimulation of TNF alpha expression and autocrine induction of p38 MAPK signaling cascade [79]. There is also evidence that protein kinase C delta can inhibit the Akt survival pathway, inducing Akt dephosphorylation [80].

Besides proliferation and apoptosis, the androgen receptor can also affect cell motility and the invasive growth of cells through regulation of the c-Met proto-oncogene (Fig. **4**).

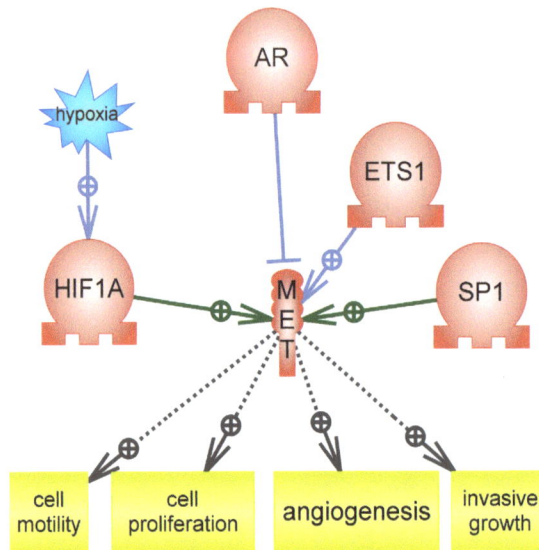

Figure 4: AR signaling through Met proto-oncogene.

The transmembrane receptor tyrosine kinase, c-Met, is activated by the hepatocyte growth factor/scatter factor (HGF/SF), a multifunctional growth factor that plays a critical role in the regulation of cell growth, cell motility, morphogenesis, and angiogenesis [81]. The aberrant expression of HGF/SF or c-Met was shown to regulate the invasion and growth of carcinoma cells in prostate cancer and correlates with poor prognosis in cancer patients [82]. Notably, the androgen receptor negatively regulates c-Met promoter: AR represses Sp1-induced c-Met promoter activity by interfering with Sp1 binding to the promoter region of c-Met. For example, castration induces the c-Met expression in LNCaP cell xenografts [83] that may lead to the development of more aggressive tumors.

The androgen receptor also maintains communication between epithelial cells and regulates gap junction formation by influencing the expression of connexins (Fig. **5**).

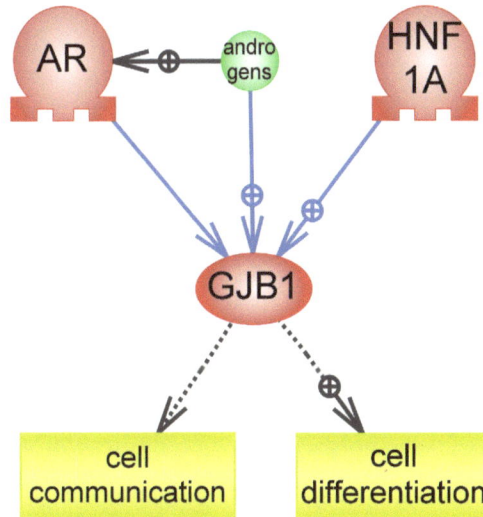

Figure 5: Androgen receptor regulation of gap junction formation and cell communication.

Connexins forming gap junctions play a significant role in communication, transport, and signaling between epithelial cells [84]. Connexin 32 (Cx32, GJB1) is also shown to be a tumor suppressor in several tissues such as the liver and the lung [85]. It was shown that the expression, trafficking, and assembly of Cx43 and Cx32 are downregulated during prostate cancer progression, whereas they expressed by the well-differentiated epithelial cells of the normal prostate [86]. Forced expression of Cx32 and Cx43 in the androgen-responsive human prostate cancer cell line decreases the rate of growth and induces differentiation [86]. Furthermore, it was shown that androgens increase Cx32 expression level and gap junction formation. There is also evidence that in the absence of androgens, Cx32 is degraded either from the ER or immediately upon its exit before reaching Golgi, presumably, *via* endoplasmic reticulum-associated degradation (ERAD) [86].

All of the previous examples provide evidence of the importance of AR in the regulation of cell proliferation, apoptosis, invasive growth and cell-to-cell communication. Given androgen-deprived conditions, impaired AR activity can lead to disturbances in these pathways, thereby influencing homeostasis of the prostate. It is generally accepted that the androgen receptor plays an important role in prostate cancer even upon transition to an androgen-independent state. This can be due to changes either in the receptor itself (*e.g.*, mutations or amplifications) or in coregulators influencing AR activity (*e.g.*, NCOA2, NCOA3,

NCOA4, cyclin D1, or ARA55) [31, 38]. But all of these alterations could probably change the transcriptional program performed by AR and lead to aberrant signaling, which is apparently one of the causes of further deregulation of cell signaling cascades that accomapny hormone-refractory status and poor prognosis.

CONCLUSION

Androgen deprivation therapy is initially an effective treatment of locally advanced prostate cancer. But in most cases ADT leads to the development of an androgen-independent, or hormone-refractory, status characterized by numerous alterations in basic signaling pathways that play a major role in cell homeostasis. As we reviewed in the section "Biochemical characterization of prostate cancer – overview pathway," many signaling cascades responsible for pro-survival, proliferative and apoptotic programs are deregulated in advanced cases of prostate cancer. Therefore, ADT can be considered partially responsible for transition to advanced stages of the disease. In the section "The role of AR in normal prostate and cancerous tissue" we have provided further evidence for this statement. We showed that the androgen receptor has a central role in prostate homeostasis. It can regulate both proliferation and apoptosis in a balanced manner but also plays a role in the metastasis potential of the prostate cells and in cell-to-cell communication. Both cancer progression and androgen deprivation therapy can disturb the balance of pro-survival and proapoptotic programs performed by a normally working androgen receptor. In summary, we suggest that androgen deprivation therapy may not be the best way to manage prostate cancer, and more effective and safer options for the treatment of this disease have to be researched.

CONFLICT OF INTEREST

Authors do not have any conflicts of interests with respect to chapter content.

ACKNOWLEDGEMENTS

We wish to thank Dr. Andrey Kalinin for his assistance in drawing cancer overview pathway and Dr. Anton Fedorov for the summarizing useful information on prostate cancer.

REFERENCES

[1]   Vindrieux D, Escobar P, Lazennec G. Emerging roles of chemokines in prostate cancer. *Endocr Relat Cancer.* 2009 Sep;16(3):663-73.

[2]   Jemal A, Bray F, Center MM, Ferlay J, Ward E, Forman D. Global cancer statistics. *CA Cancer J Clin.* 2011 Mar-Apr;61(2):69-90.

[3]   Bruner DW, Moore D, Parlanti A, Dorgan J, Engstrom P. Relative risk of prostate cancer for men with affected relatives: systematic review and meta-analysis. *Int J Cancer.* 2003 Dec 10;107(5):797-803.

[4]   Lapointe J, Li C, Higgins JP, van de Rijn M, Bair E, Montgomery K, *et al.,* Gene expression profiling identifies clinically relevant subtypes of prostate cancer. *Proc Natl Acad Sci U S A.* 2004 Jan 20;101(3):811-6.

[5]   Datta K, Muders M, Zhang H, Tindall DJ. Mechanism of lymph node metastasis in prostate cancer. *Future Oncol.* 2010 May;6(5):823-36.

[6]   Rentsch CA, Cecchini MG, Thalmann GN. Loss of inhibition over master pathways of bone mass regulation results in osteosclerotic bone metastases in prostate cancer. *Swiss Med Wkly.* 2009 Apr 18;139(15-16):220-5.

[7]   Catton C. The role of radiation therapy in prostate cancer after radical prostatectomy: when and why? *Curr Opin Support Palliat Care.* 2010 Sep;4(3):135-40.

[8]   Singh J, Trabulsi EJ, Gomella LG. Is there an optimal management for localized prostate cancer? *Clin Interv Aging.* 2010 Aug 9;5:187-97.

[9]   Nassif AE, Tâmbara Filho R. Immunohistochemistry expression of tumor markers CD34 and P27 as a prognostic factor of clinically localized prostate adenocarcinoma after radical prostatectomy. *Rev Col Bras Cir.* 2010 Oct;37(5):338-44.

[10] Sharifi N, Gulley JL, Dahut WL. An update on androgen deprivation therapy for prostate cancer. *Endocr Relat Cancer.* 2010 Oct 29;17(4):R305-15.

[11] Koochekpour S. Androgen receptor signaling and mutations in prostate cancer. *Asian J Androl.* 2010 Sep;12(5):639-57.

[12] Mostaghel EA, Montgomery B, Nelson PS. Castration-resistant prostate cancer: targeting androgen metabolic pathways in recurrent disease. *Urol Oncol.* 2009 May-Jun; 27(3):251-7.

[13] Fielding LP, Genglio-Presier CM, Freedman LS. The future of prognostic factors in outcome prediction for patients with cancer. *Cancer.* 1992; 70: 2367-77.

[14] American Joint Committee on Cancer. Prostrate: in *AJCC Cancer Staging Manual*, 6th ed. New York, NY: Springer. 2002: 309-16.

[15] Mellinger GT, Gleason D, Bailar J 3rd. The histology and prognosis of prostatic cancer. *J Urol.* 1967 Feb;97(2):331-7.

[16] Epstein JI. An update of the Gleason grading system. *J Urol.* 2010 Feb; 183(2):433-40.

[17] Partin AW, Yoo J, Carter HB, *et al.,* The use of prostate specific antigen, clinical stage and Gleason score to predict pathological stage in men with localized prostate cancer. J *Urol.* 1993; 150: 110-4.

[18] D'Amico AV, Whittington R, Malkowicz SB, *et al.,* Biochemical outcome after radical prostatectomy, external beam radiation therapy, or interstitial radiation therapy for clinically localized prostate cancer. *JAMA.* 1998; 280:969–974.

[19] Pomerantz M, Kantoff P. Advances in the treatment of prostate cancer. *Annu Rev Med.* 2007; 58:205-20.

[20] Hummel S, Simpson EL, Hemingway P, Stevenson MD, Rees A. Intensity-modulated radiotherapy for the treatment of prostate cancer: a systematic review and economic evaluation. *Health Technol Assess.* 2010 Oct;14(47):1-108, iii-iv.

[21] Rosser CJ, Gaar M, Porvasnik S. Molecular fingerprinting of radiation resistant tumors: can we apprehend and rehabilitate the suspects? *BMC Cancer.* 2009 Jul 9;9:225.

[22] [Internet] National Cancer Institute, http://www.cancer.gov/cancertopics/pdq/treatment/ prostate/HealthProfessional/page4.

[23] Madu CO, Lu Y. Novel diagnostic biomarkers for prostate cancer. *J Cancer.* 2010 Oct 6;1:150-77.

[24] [Internet] Prostate Cancer Foundation http://www.pcf.org/site/c.leJRIROrEpH/b.5813295/ k.FA2E/Active_Surveillance.htm

[25] Fang LC, Merrick GS, Wallner KE. Androgen deprivation therapy: a survival benefit or detriment in men with high-risk prostate cancer? *Oncology* (Williston Park). 2010 Aug; 24(9):790-6, 798.

[26] Sharifi N, Gulley JL, Dahut WL. Androgen deprivation therapy for prostate cancer. *JAMA.* 2005 Jul 13;294(2):238-44.

[27] Locke JA, Guns ES, Lubik AA, Adomat HH, Hendy SC, Wood CA, Ettinger SL, Gleave ME, Nelson CC. Androgen levels increase by intratumoral *de novo* steroidogenesis during progression of castration-resistant prostate cancer. *Cancer Res.* 2008 Aug 1;68(15):6407-15.

[28] Shahinian VB, Kuo YF, Freeman JL, Orihuela E, Goodwin JS. Characteristics of urologists predict the use of androgen deprivation therapy for prostate cancer. *J Clin Oncol.* 2007 Dec 1;25(34):5359-65.

[29] Nwosu V, Carpten J, Trent JM, Sheridan R. Heterogeneity of genetic alterations in prostate cancer: evidence of the complex nature of the disease. *Hum Mol Genet.* 2001 Oct 1;10(20):2313-8.

[30] Lee JT, Lehmann BD, Terrian DM, Chappell WH, Stivala F, Libra M, Martelli AM, Steelman LS, McCubrey JA. Targeting prostate cancer based on signal transduction and cell cycle pathways. *Cell Cycle.* 2008 Jun 15;7(12):1745-62.

[31] Rahman M, Miyamoto H, Chang C. Androgen receptor coregulators in prostate cancer: mechanisms and clinical implications. *Clin Cancer Res.* 2004 Apr 1;10(7):2208-19.

[32] Lin HK, Yeh S, Kang HY, Chang C. Akt suppresses androgen-induced apoptosis by phosphorylating and inhibiting androgen receptor. *Proc Natl Acad Sci U S A.* 2001 Jun 19;98(13):7200-5.

[33] Balk SP, Knudsen KE. AR, the cell cycle, and prostate cancer. *Nucl Recept Signal.* 2008 Feb 1;6:e001.

[34] Lei Q, Jiao J, Xin L, Chang CJ, Wang S, Gao J, Gleave ME, Witte ON, Liu X, Wu H. NKX3.1 stabilizes p53, inhibits AKT activation, and blocks prostate cancer initiation caused by PTEN loss. *Cancer Cell.* 2006 May;9(5):367-78.

[35] Cha TL, Zhou BP, Xia W, Wu Y, Yang CC, Chen CT, Ping B, Otte AP, Hung MC. Akt-mediated phosphorylation of EZH2 suppresses methylation of lysine 27 in histone H3. *Science.* 2005 Oct 14;310(5746):306-10.

[36] Papatsoris AG, Karamouzis MV, Papavassiliou AG. The power and promise of "rewiring" the mitogen-activated protein kinase network in prostate cancer therapeutics. Mol Cancer Ther. 2007 Mar;6(3):811-9.

[37]   Neto AS, Tobias-Machado M, Wroclawski ML, Fonseca FL, Teixeira GK, Amarante RD, Wroclawski ER, Del Giglio A. Her-2/neu expression in prostate adenocarcinoma: a systematic review and meta-analysis. *J Urol.* 2010 Sep;184(3):842-50.

[38]   Heinlein CA, Chang C. Androgen receptor in prostate cancer. *Endocr Rev.* 2004 Apr;25(2):276-308.

[39]   Chen T, Wang LH, Farrar WL. Interleukin 6 activates androgen receptor-mediated gene expression through a signal transducer and activator of transcription 3-dependent pathway in LNCaP prostate cancer cells. *Cancer Res.* 2000 Apr 15; 60(8):2132-5.

[40]   Feng XH, Lin X, Derynck R. Smad2, Smad3 and Smad4 cooperate with Sp1 to induce p15(Ink4B) transcription in response to TGF-beta. *EMBO J.* 2000 Oct 2; 19(19):5178-93.

[41]   Liu SK, Hoffmann FM. Smad4 cooperates with lymphoid enhancer-binding factor 1/T cell-specific factor to increase c-myc expression in the absence of TGF-beta signaling. *Proc Natl Acad Sci U S A.* 2006 Dec 5;103(49):18580-5.

[42]   Yuan TC, Veeramani S, Lin MF. Neuroendocrine-like prostate cancer cells: neuroendocrine transdifferentiation of prostate adenocarcinoma cells. *Endocr Relat Cancer.* 2007 Sep;14(3):531-47.

[43]   Sun Y, Niu J, Huang J. Neuroendocrine differentiation in prostate cancer. *Am J Transl Res.* 2009 Feb 5;1(2):148-62.

[44]   Honer M, Mu L, Stellfeld T, Graham K, Martic M, Fischer CR, Lehmann L, Schubiger PA, Ametamey SM, Dinkelborg L, Srinivasan A, Borkowski S. 18F-Labeled Bombesin Analog for Specific and Effective Targeting of Prostate Tumors Expressing Gastrin-Releasing Peptide Receptors. *J Nucl Med.* 2011 Jan 13.

[45]   Pang Y, Young CY, Yuan H. MicroRNAs and prostate cancer. *Acta Biochim Biophys Sin* (Shanghai). 2010 Jun 15;42(6):363-9.

[46]   Saini S, Majid S, Dahiya R. Diet, microRNAs and prostate cancer. *Pharm Res.* 2010 Jun;27 (6):1014-26.

[47]   Huggins C, Hodges CV. Studies on prostatic cancer, I: the effect of estrogen and of androgen injection on serum phosphatases in metastatic carcinoma of the prostate. *Cancer Res.* 1941;1:293-297.

[48]   Yeh S, Tsai MY, Xu Q, Mu XM, Lardy H, Huang KE, Lin H, Yeh SD, Altuwaijri S, Zhou X, Xing L, Boyce BF, Hung MC, Zhang S, Gan L, Chang C Generation and characterization of androgen receptor knockout (ARKO) mice: an *in vivo* model for the study of androgen functions in selective tissues. *Proc Natl Acad Sci USA* 2002 (99):13498–13503.

[49]   Dell'Edera D, Malvasi A, Vitullo E, Epifania AA, Tinelli A, Laterza F, Novelli A, Pacella E, Mazzone E, Novelli G. Androgen insensitivity syndrome (or Morris syndrome) and other associated pathologies. *Eur Rev Med Pharmacol Sci.* 2010 Nov;14(11):947-57.

[50]   English HF, Kyprianou N, Isaacs JT 1989 Relationship between DNA fragmentation and apoptosis in the programmed cell death in the rat prostate following castration. *Prostate* 15:233–250.

[51]   Michael IP, Pampalakis G, Mikolajczyk SD, Malm J, Sotiropoulou G, Diamandis EP. Human tissue kallikrein 5 is a member of a proteolytic cascade pathway involved in seminal clot liquefaction and potentially in prostate cancer progression. *J Biol Chem.* 2006 May 5;281(18):12743-50.

[52]   Golovine K, Makhov P, Uzzo RG, Shaw T, Kunkle D, Kolenko VM. Overexpression of the zinc uptake transporter hZIP1 inhibits nuclear factor-kappaB and reduces the malignant

potential of prostate cancer cells *in vitro* and *in vivo*. *Clin Cancer Res.* 2008 Sep 1;14(17):5376-84.

[53]   Franklin RB, Costello LC. Zinc as an anti-tumor agent in prostate cancer and in other cancers. *Arch Biochem Biophys.* 2007 Jul 15;463(2):211-7.

[54]   Song H, Zhang B, Watson MA, Humphrey PA, Lim H, Milbrandt J. Loss of Nkx3.1 leads to the activation of discrete downstream target genes during prostate tumorigenesis. *Oncogene.* 2009 Sep 17;28(37):3307-19.

[55]   Lei Q, Jiao J, Xin L, Chang CJ, Wang S, Gao J, Gleave ME, Witte ON, Liu X, Wu H. NKX3.1 stabilizes p53, inhibits AKT activation, and blocks prostate cancer initiation caused by PTEN loss. *Cancer Cell.* 2006 May;9(5):367-78.

[56]   Muhlbradt E, Asatiani E, Ortner E, Wang A, Gelmann EP. NKX3.1 activates expression of insulin-like growth factor binding protein-3 to mediate insulin-like growth factor-I signaling and cell proliferation. *Cancer Res.* 2009 Mar 15;69(6):2615-22.

[57]   Ouyang X, DeWeese TL, Nelson WG, Abate-Shen C. Loss-of-function of Nkx3.1 promotes increased oxidative damage in prostate carcinogenesis. *Cancer Res.* 2005 Aug 1;65(15):6773-9.

[58]   Li X, Guan B, Maghami S, Bieberich CJ. NKX3.1 is regulated by protein kinase CK2 in prostate tumor cells. *Mol Cell Biol.* 2006 Apr;26(8):3008-17.

[59]   Markowski MC, Bowen C, Gelmann EP. Inflammatory cytokines induce phosphorylation and ubiquitination of prostate suppressor protein NKX3.1. *Cancer Res.* 2008 Sep 1;68(17):6896-901.

[60]   Romano S, Di Pace A, Sorrentino A, Bisogni R, Sivero L, Romano MF. FK506 binding proteins as targets in anticancer therapy. *Anticancer Agents Med Chem.* 2010 Nov 1;10(9):651-6.

[61]   Sinars CR, Cheung-Flynn J, Rimerman RA, Scammell JG, Smith DF, Clardy J. Structure of the large FK506-binding protein FKBP51, an Hsp90-binding protein and a component of steroid receptor complexes. *Proc Natl Acad Sci U S A.* 2003 Feb 4;100(3):868-73.

[62]   Tatro ET, Everall IP, Kaul M, Achim CL. Modulation of glucocorticoid receptor nuclear translocation in neurons by immunophilins FKBP51 and FKBP52: implications for major depressive disorder. *Brain Res.* 2009 Aug 25;1286:1-12.

[63]   De Bosscher K, Vanden Berghe W, Haegeman G. The interplay between the glucocorticoid receptor and nuclear factor-kappaB or activator protein-1: molecular mechanisms for gene repression. *Endocr Rev.* 2003 Aug;24(4):488-522.

[64]   Yemelyanov A, Czwornog J, Chebotaev D, Karseladze A, Kulevitch E, Yang X, Budunova I. Tumor suppressor activity of glucocorticoid receptor in the prostate. *Oncogene.* 2007 Mar 22;26(13):1885-96.

[65]   Huang WC, Wu D, Xie Z, Zhau HE, Nomura T, Zayzafoon M, Pohl J, Hsieh CL, Weitzmann MN, Farach-Carson MC, Chung LW. beta2-microglobulin is a signaling and growth-promoting factor for human prostate cancer bone metastasis. *Cancer Res.* 2006 Sep 15;66(18):9108-16.

[66]   Yoon HG, Wong J. The corepressors silencing mediator of retinoid and thyroid hormone receptor and nuclear receptor corepressor are involved in agonist- and antagonist-regulated transcription by androgen receptor. *Mol Endocrinol.* 2006 May;20(5):1048-60.

[67]   Nomura T, Huang WC, Zhau HE, Wu D, Xie Z, Mimata H, Zayzafoon M, Young AN, Marshall FF, Weitzmann MN, Chung LW. Beta2-microglobulin promotes the growth of human renal cell carcinoma through the activation of the protein kinase A, cyclic AMP-

responsive element-binding protein, and vascular endothelial growth factor axis. *Clin Cancer Res.* 2006 Dec 15;12(24):7294-305.

[68] King WG, Mattaliano MD, Chan TO, Tsichlis PN, Brugge JS. Phosphatidylinositol 3-kinase is required for integrin-stimulated AKT and Raf-1/mitogen-activated protein kinase pathway activation. *Mol Cell Biol.* 1997 Aug;17(8):4406-18.

[69] Franke TF, Kaplan DR, Cantley LC, Toker A. Direct regulation of the Akt proto-oncogene product by phosphatidylinositol-3,4-bisphosphate. *Science.* 1997 Jan 31;275(5300):665-8.

[70] Fedele CG, Ooms LM, Ho M, Vieusseux J, O'Toole SA, Millar EK. et. al. Inositol polyphosphate 4-phosphatase II regulates PI3K/Akt signaling and is lost in human basal-like breast cancers. *Proc Natl Acad Sci U S A.* 2010 Dec 21;107(51):22231-6.

[71] Hodgson MC, Shao LJ, Frolov A, Li R, Peterson LE, Ayala G, Ittmann MM, Weigel NL, Agoulnik IU. Decreased expression and androgen regulation of the tumor suppressor gene INPP4B in prostate cancer. *Cancer Res.* 2011 Jan 15;71(2):572-82. Epub 2011 Jan 11.

[72] Sherk AB, Frigo DE, Schnackenberg CG, Bray JD, Laping NJ, Trizna W, Hammond M, Patterson JR, Thompson SK, Kazmin D, Norris JD, McDonnell DP. Development of a small-molecule serum- and glucocorticoid-regulated kinase-1 antagonist and its evaluation as a prostate cancer therapeutic. *Cancer Res.* 2008 Sep 15;68(18):7475-83.

[73] Shanmugam I, Cheng G, Terranova PF, Thrasher JB, Thomas CP, Li B. Serum/glucocorticoid-induced protein kinase-1 facilitates androgen receptor-dependent cell survival. *Cell Death Differ.* 2007 Dec;14(12):2085-94.

[74] Zhang BH, Tang ED, Zhu T, Greenberg ME, Vojtek AB, Guan KL. Serum- and glucocorticoid-inducible kinase SGK phosphorylates and negatively regulates B-Raf. *J Biol Chem.* 2001 Aug 24;276(34):31620-6.

[75] Aoyama T, Matsui T, Novikov M, Park J, Hemmings B, Rosenzweig A. Serum and glucocorticoid-responsive kinase-1 regulates cardiomyocyte survival and hypertrophic response. *Circulation.* 2005 Apr 5;111(13):1652-9.

[76] Rajamanickam J, Palmada M, Lang F, Boehmer C. EAAT4 phosphorylation at the SGK1 consensus site is required for transport modulation by the kinase. *J Neurochem.* 2007 Aug;102(3):858-66.

[77] Reyland ME. Protein kinase C delta and apoptosis. Biochem Soc Trans. 2007 Nov;35(Pt 5):1001-4.

[78] Gavrielides MV, Gonzalez-Guerrico AM, Riobo NA, Kazanietz MG. Androgens regulate protein kinase Cdelta transcription and modulate its apoptotic function in prostate cancer cells. *Cancer Res.* 2006 Dec 15;66(24):11792-801.

[79] Kilpatrick LE, Sun S, Li H, Vary TC, Korchak HM. Regulation of TNF-induced oxygen radical production in human neutrophils: role of delta-PKC. *J Leukoc Biol.* 2010 Jan;87(1):153-64.

[80] Dubi N, Gheber L, Fishman D, Sekler I, Hershfinkel M. Extracellular zinc and zinc-citrate, acting through a putative zinc-sensing receptor, regulate growth and survival of prostate cancer cells. *Carcinogenesis.* 2008 Sep;29(9):1692-700.

[81] Lesko E, Majka M. The biological role of HGF-MET axis in tumor growth and development of metastasis. *Front Biosci.* 2008 Jan 1;13:1271-80.

[82] Niu YN, Xia SJ. Stroma-epithelium crosstalk in prostate cancer. *Asian J Androl.* 2009 Jan;11(1):28-35.

[83] Verras M, Lee J, Xue H, Li TH, Wang Y, Sun Z. The androgen receptor negatively regulates the expression of c-Met: implications for a novel mechanism of prostate cancer progression. *Cancer Res.* 2007 Feb 1; 67(3):967-75.

[84]   Laird DW. Life cycle of connexins in health and disease. *Biochem J.* 2006 Mar 15;394(Pt 3):527-43.

[85]   Zhang W, Li HG, Fan MJ, Lv ZQ, Shen XM, He XX. Expressions of connexin 32 and 26 and their correlation to prognosis of non-small cell lung cancer. *Ai Zheng.* 2009 Feb;28(2):173-6.

[86]   Mitra S, Annamalai L, Chakraborty S, Johnson K, Song XH, Batra SK, Mehta PP. Androgen-regulated formation and degradation of gap junctions in androgen-responsive human prostate cancer cells. *Mol Biol Cell.* 2006 Dec;17(12):5400.

CHAPTER 2

Role of Ca^{2+}-Mediated Signaling in ALS Pathology

Ekaterina A. Kotelnikova[1,*], Mikhail A. Pyatnitskiy[1], Rachel L. Redler[2] and Nikolay V. Dokholyan[2]

[1]Ariadne Genomics Inc., Rockville, MD USA and [2]University of North Carolina, Chapel Hill, NC USA

Abstract: Familial amyotrophic lateral sclerosis (fALS) is a hereditary disorder of motor neurons that is caused by mutation in Cu, Zn superoxide dismutase (SOD1) in a subset of cases. The onset of the disease is relatively late, usually at age 50 or later, and is associated with interrelated molecular mechanisms of neurodegeneration. One of the mechanisms that can promote ALS progression is increased intracellular calcium concentration. The only market-available drug for ALS targets glutamate receptors and slows disease in part by mitigating excitotoxicity, a process in which persistent stimulation of glutamate receptors leads to pathologically high calcium concentration. To dissect the potential contributions of calcium mishandling to ALS, we have processed several publically available expression datasets related to fALS and analyzed the differential expression of genes related to calcium homeostasis. We find that SOD1-related fALS is associated with changes in expression of numerous genes related to calcium handling. Several genes which are down-regulated in fALS are targets of the repressor element-1 transcription factor/neuron restrictive silencer factor (REST/NRSF) transcription factor, which is normally inactivated in neuronal tissue. Our meta-analysis shows that changes in gene expression occurring in SOD1-related fALS promote calcium mishandling through dysregulation of multiple pathways, and that aberrant REST/NRSF activity may underlie some errors in calcium homeostasis.

Keywords: Amyotrophic lateral sclerosis, ALS, Lou Gehrig's disease, Bulbar Motor Neuron Disease, bioinformatics, pathway analysis, Pathway Studio, disease mechanism, gene expression microarray, SOD1 mice, calcium signaling, REST transcription factor.

INTRODUCTION

Amyotrophic lateral sclerosis (ALS) is a late-onset neurodegenerative disorder, in which selective death of motor neurons causes progressive paralysis and death, typically within 3 years [1]. It is known that loss of motor neurons in amyotrophic

*Address correspondence to Ekaterina A. Kotelnikova: Ariadne Genomics Inc., Rockville, USA; E-mail: ekotelnikova@ariadnegenomics.com

lateral sclerosis (ALS) is associated with several processes, including increased intracellular calcium, glutamate excitotoxicity, mitochondrial dysfunction, oxidative stress, neurofilament aggregation and dysfunction of axonal transport mechanisms [1]. The most prominent genetic factor in familial ALS (fALS) is mutation of the gene encoding Cu, Zn superoxide dismutase (SOD1), which accounts for 20% of fALS cases and is the basis for most animal models of the disease [2]. A key attribute of most cases of SOD1-linked fALS is abnormal aggregation of mutant SOD1 and subsequent motor neuron death [3], but the primary trigger(s) leading to the onset of neurodegeneration are still under question. Dysregulation of intracellular calcium concentration is a potentially important factor in ALS pathogenesis, as evidenced by the ability of elevated $[Ca^{2+}]$ to instigate many other cellular dysfunctions linked to the disease [1]. However, the role of dysregulation of calcium signaling in ALS has not yet been extensively explored. Here we analyze microarray expression data of ALS patients and mouse models from the Gene Expression Omnibus (GEO) database in order to clarify the mechanisms of calcium signaling in the context of the ALS disease state.

RESULTS

The analysis of differential expression values from spinal cord tissues of familial ALS patients *vs.* non-neurological controls and of spinal cord and muscle from SOD1-G93A transgenic mice *vs.* transgenic mice expressing wild type SOD1 using the SNEA (Sub-Network Enrichment Analysis) algorithm with the Mann-Whitney test shows that various sub-networks associated with inflammation and immunity-related processes are activated. Another common feature of tissue samples from either ALS patients or SOD1-G93A mice is the statistically significant overrepresentation of differentially expressed genes related to calcium signaling (Supplementary Tables **1** and **2**). Inflammation has long been associated with the pathoprogression of ALS [4, 5]. However, the inflammatory response seems to be consequent to mutant SOD1 aggregation, which is the key attribute of most cases of fALS. In contrast, the intracellular calcium concentration has been discussed extensively as a possible primary trigger of the disease [6–8]. Taking the hypothesis that calcium concentration is one of the primary pathological triggers which initiate motor neuron dysfunction in fALS, we analyzed available microarray data to distinguish the contribution of individual gene expression changes to this potentially important mechanism.

Literature Overview of Calcium's Role in fALS

In order to analyze genes which are potentially involved in calcium dysregulation in ALS, we first constructed a literature-based overview pathway of calcium signaling in ALS-affected motor neurons (Fig. **1**). We combined the information about gene regulation in this pathway with differential expression values (log-ratio with base 2 outside [-1,1] interval) between familial ALS (fALS) and normal individuals, and between SOD1-G93A mice and SOD1-WT mice (Supplementary Table **3**).

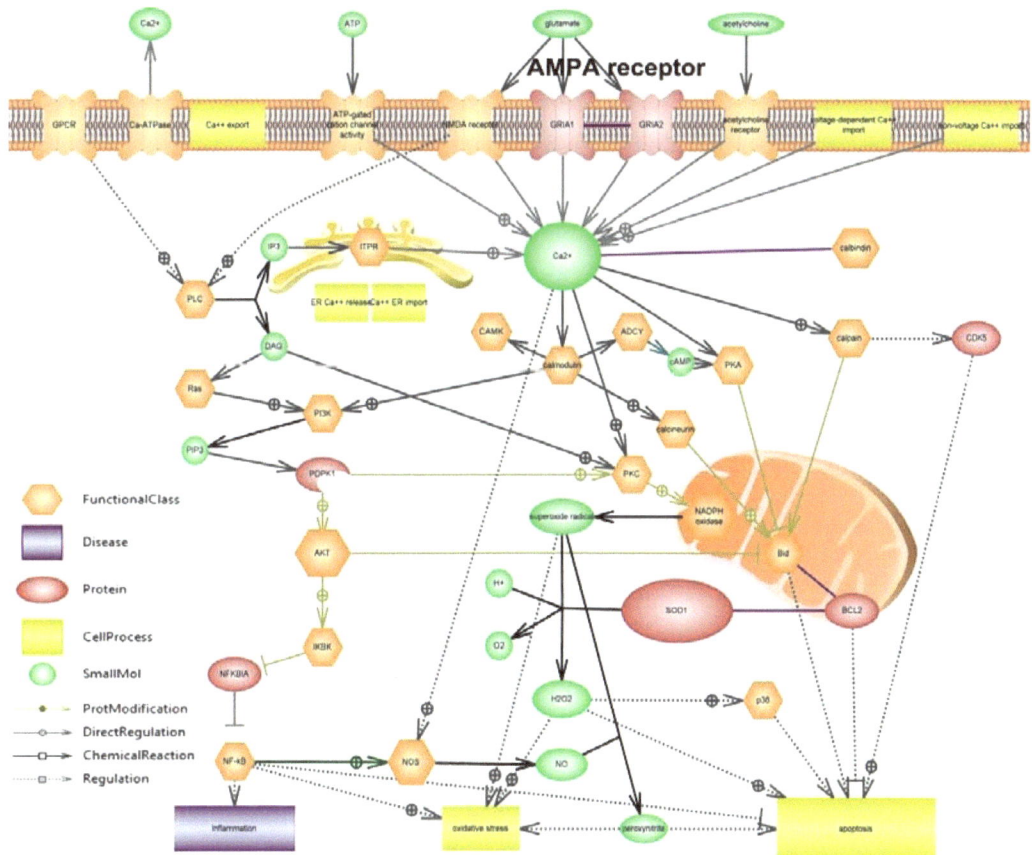

Figure 1: Calcium signaling in SOD1-dependent ALS. Functional Class entities represent several protein members all performing the same function within the context of this pathway. Cell Process entities contain several protein members directly involved in the same process.

We find that fALS alters expression of several genes that directly regulate calcium homeostasis (Supplementary Table **3**) in muscle and/or motor neurons, with the common consequence of increased cytosolic calcium. These genes do not have an "indirect" or "downstream" mark in Supplementary Table **3** and are known to directly affect calcium concentration in various tissues.

Dysregulation of proteins related to calcium buffering, storage and export. ALS-vulnerable motor neurons express lower levels of calcium-binding proteins compared to resistant neural populations such as those that control eye movement. The diminished calcium-buffering capacity of motor neurons, while essential for their function, may also contribute to their selective vulnerability in ALS [9–11]. At least one of these cytosolic calcium-binding proteins, calbindin (CALB1, a member of the troponin C superfamily and of the "calbindin" functional class in Fig. **1**), is down-regulated in fALS, further diminishing cellular calcium buffering capacity.

Cytosolic calcium concentration may also increase in fALS-affected tissue due to changes in expression of proteins that regulate transport of calcium into "sinks" such as the endoplasmic and sarcoplasmic reticula, or into the extracellular space. fALS patients and mouse models show decreased expression of the gene encoding SERCA Ca^{2+}-ATPase (ATP2A3, a member of the "Ca^{++} ER import" cell process and of the "Ca-ATPase" functional class shown in Fig. **1**), which facilitates Ca^{2+} import into the endoplasmic reticulum (ER) and sarcoplasmic reticulum (SR). ATP2A3 expression also declines in rat cerebral cortex during natural aging [12], giving aged neurons additional vulnerability to excess cytosolic Ca^{2+} that may contribute to the late-onset nature of ALS. Intracellular calcium is also increased in fALS through upregulation of ITPR2 (member of the "ER Ca^{++} release" cell process and of the "ITPR" functional class in Fig. **1**), which mediates release of Ca^{2+} from the ER in response to the second messenger inositol 1,4,5-trisphosphate. Interestingly, variants in ITPR2 that result in increased expression have been linked to a subset of sporadic ALS cases [13]. Another Ca^{2+}-ATPase, the plasma membrane protein ATP2B2 (a member of the "Ca^{++} ER import" cell process and of the "Ca-ATPase" functional class in Fig. **1**), is also down-regulated in fALS, resulting in a diminished capacity to export Ca^{2+} from the cytosol.

fALS also induces changes in expression of the ryanodine receptor isoform RYR2, which mediates calcium-induced calcium release from the ER in muscle and the nervous system [14,15]. Also apparent in fALS are down-regulation of CACNA1C and strong up-regulation of CACNB3, two subunits of voltage-gated L-type calcium channels. It is known that most patients with amyotrophic lateral sclerosis possess antibodies (ALS IgGs) that bind to L-type skeletal muscle voltage-gated calcium channels and inhibit L-type calcium current [16–18]. The role of differential expression of L-type calcium channel subunits in ALS autoimmunity should be further evaluated.

Defective glutamate signaling. Glutamate, an excitatory amino acid, activates different types of ion channel-forming (ionotropic) receptors to develop their role in neurons. The classification of ionotropic glutamate receptors is based on their activation by different pharmacologic agonists: AMPA (a-amino-3-hydroxy-5-methyl-4-isoxazolepropionic acid), kainate, and NMDA (N-methyl-D-aspartic acid). AMPA and kainate receptors trigger rapid excitatory neurotransmission in the central nervous system (CNS) by promoting entry of Na^+ into neurons, but can also be permeable to Ca^{2+}. The GluR2 subunit encoded by the GRIA2 gene is of particular functional significance because of its effect on calcium permeability. Most AMPA receptors in the human CNS include the edited GluR2 subunit, GluR2(R), making them impermeability to calcium [11,19]. The low expression of GluR2 by human motor neurons implies that most of the surface AMPA receptors expressed by this cell group are likely to be atypical and calcium permeable. The down-regulation of GRIA2 expression in motor neurons of fALS patients and mutant SOD1 mice would increase Ca^{2+} influx by increasing the calcium permeability of AMPA receptors. GRIA2 down-regulation may be an important pathological mechanism in fALS since Ca^{2+}-influx through atypical motor neuronal AMPA receptors can promote misfolding of mutant SOD1 protein and eventual death of these neurons [20].

Expression of NMDA receptor components is also perturbed in fALS through down-regulation of GRIN1, which encodes the NR1 subunit of the NMDA receptor, as well as GRIN2A and GRIN2C, which encode the NR2 subunits. NMDA receptor activation, which requires binding of glutamate and glycine, also leads to an influx of Ca^{2+} into the postsynaptic region. NMDA and AMPA receptors are known to be implicated in glutamate excitotoxicity, which is considered to be a prominent contributor to ALS progression [20,21]. The significant widespread loss of the NR2A subunit was also shown elsewhere for ALS as compared to controls [22], but how this down-regulation could lead to increased excitotoxicity remains unclear.

In summary, the above data shows that changes in gene expression that occur in fALS may lead to disturbance in Ca^{2+} export from the cytoplasm (Ca^{2+}-ATPases ATP2A3/ATP2B2), Ca^{2+} buffering (calbindin) and Ca^{2+} transportation from the ER (ITPR2), as well as induction of the well-known mechanism of excitotoxicity.

Genes linked to the calcium-related processes. To further investigate possible mechanisms of calcium deregulation in ALS we generated calcium-related sub-networks using Sub-Network Enrichment Analysis (SNEA) using Cell Processes as downstream seeds. Sub-networks were generated using: a) all log-ratios from fALS vs normal tissue (using the Mann-Whitney test); b) all log-ratios from SOD1-G93A mice *vs.* SOD1-WT mice (using the Mann-Whitney test) and c) differentially expressed genes with combined rank smaller than 1000 (see Materials and Methods) from SOD1-G93A mice *vs.* SOD1-WT mice (using Fischer's exact test). In the resulting sub-networks, only genes connected to "Amyotrophic Lateral Sclerosis" in the ResNet database and having a log-ratio value with absolute value greater than 1 and the same sign for both human and mice experiments were retained for visualization and biological interpretation in Fig. **2**.

We found that many genes from Fig. **2** contributing to calcium regulation are involved in inflammation and immunity processes (C5AR1, CCL13, CCL5, CCR1, CD72, CD9, CD97, CX3CL1, CXCR4, FCER1G, FCGR1A, FCGR2B, GZMB, ICAM1, ITGAL, ITGAM, ITGB2, KLRK1, LGALS3, NT5E, SELL, WAS, ZAP70). Also implicated is TNNT2, troponin T type 2, which is relevant only to processes occuring in muscle. To focus on primary triggers of calcium mishandling in motor neurons, we remove from consideration genes whose expression is limited to muscle or whose involvement is limited to inflammatory or immune responses (late-occurring processes in ALS pathobiology). We searched the literature for the remaining genes from Fig. **2** in order to identify those that directly or indirectly influence calcium concentration (Supplementary Table **4**).

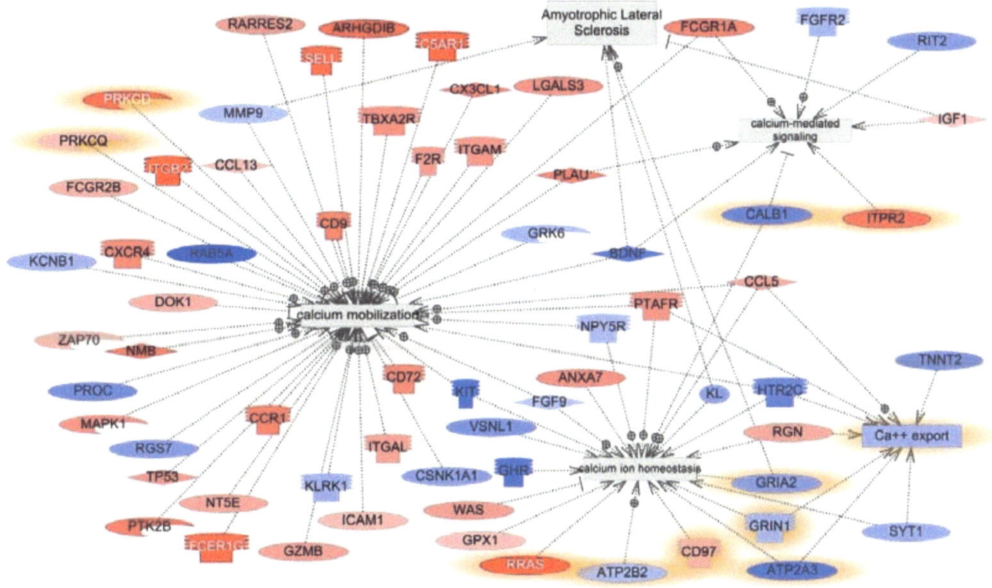

Figure 2: Calcium-related processes and genes. Red color – up-regulated genes in both fALS and in SOD1-mutant mice; blue color – down-regulated genes in both fALS and in SOD1-mutant mice. Orange highlighting – entities present in the calcium overview pathway (including members of Functional Class entities).

Direct Regulators of Calcium Concentration: Most genes directly regulating calcium concentration and differentially expressed in the ALS disease state are genes present in the overview pathway (Fig. 1): ATP2A3, ATP2B2, CALB1, CALB1, CALB1, GRIA2, GRIN1, and ITPR2. In addition, Annexin VII (ANXA7) and Regucalcin (RGN, senescence marker protein-30) are found to be up-regulated.

ANXA7 is a member of the annexin family of calcium-dependent phospholipid binding proteins. It is suggested that ANXA7 is a membrane binding protein with diverse properties, including voltage-sensitive calcium channel activity, ion selectivity and membrane fusion. It has been proposed that ANXA7 mobilizes Ca^{2+} from an endoplasmic reticulum-like pool, which can be recruited to enhance IP3-mediated Ca^{2+} release [23]. Its overexpression can thus directly or indirectly influence Ca^{2+} permeability of the motor neuron membrane.

The protein encoded by the RGN gene is a highly conserved calcium-binding protein that may have an important role in calcium homeostasis. Studies in rat indicate that this protein may also play a role in aging, as it shows age-associated down-regulation [24]. It has been shown that regucalcin is expressed in the brain, and that it can uniquely inhibit Ca^{2+}-ATPase activity in the brain microsomes of rats [25]. RGN overexpression and the resultant inhibition of calcium ATP-ases would thus result in disruption of calcium metabolism.

Indirect Regulators of Calcium Concentration: Among the differently expressed genes which indirectly affect intracellular calcium concentration are two survival factors which have been explored as potential treatments for amyotrophic lateral sclerosis: IGF1 and BDNF, whose expression is increased and decreased, respectively.

It been shown that viral administration of insulin-like growth factor 1 (IGF-1) delays ALS disease progression in the mutant SOD1 mouse model of fALS by activation of the Akt signaling pathway [26], and IGF-1 is currently undergoing clinical trials to evaluate its safety and efficacy as an ALS therapeutic [27]. IGF-1 also modulates voltage-dependent Ca^{2+} channels in neuronal cells [28]; thus, its neuroprotective properties may be a result of its effect on calcium handling. It is not clear whether up-regulation of IGF-1 in fALS is a primary event, or whether it reflects a cellular compensatory response to neuronal stressors. Brain-derived neurotrophic factor (BDNF) is known to promote motor neuron survival by activating the AKT/PI 3-kinase pathway [29]. Interestingly, BDNF induces transient calcium influx into neurons by upregulating NMDA glutamate receptor components, thus acting as modulator of excitotoxicity [30].

The other neighbors of "Amyotrophic Lateral Sclerosis" in Fig. **2** are directly connected to neuronal excitotoxicity-related proteins: GRIA2, MMP9 and Kl.

Other differentially expressed genes from Fig. **2** which may lead to intracellular calcium concentration changes in ALS motor neurons are:

- Overexpressed PTAFR (platelet-activating factor receptor). It has been shown that activation of PTAFR in transformed neuronal cell lines leads to increased intracellular calcium [31]. There are indications that PAFR-expressing neurons may be preferentially susceptible to excitotoxic challenge [32].

- Overexpressed RRAS. It has been shown that R-Ras alters Ca^{2+} homeostasis by increasing Ca^{2+} leak across the endoplasmic reticular membrane [33].

- Underexpressed CSNK1A1 (casein kinase 1, alpha 1) It was found that knockout of CK1alpha gene increases calcium mobilization [34].

- Underexpressed GRK6 (G protein-coupled receptor kinase 6). It was found that knockout of GRK6 increases calcium mobilization [34]

- Underexpressed RGS7 (regulator of G-protein signaling 7). It has been reported that RGS7 inhibited Galpha-q -coupled calcium mobilization [35,36].

Regulators of calcium-related gene expression. In order to decipher changes in transcriptional regulation which could account for the disturbed calcium homeostasis in fALS motor neurons, we selected consistently differentially expressed genes from the fALS calcium overview pathway (genes in Supplementary Table **3** without "indirect or downstream" mark) and those connected to calcium-related processes with possible function in motor neuron (genes in Supplementary Table **4**, "+" or "?" in "Possible function in motor neuron" field). We then performed a search of expression sub-networks, enriched with genes from this group, using implemented in Pathway Studio SNEA algorithm with Fischer's exact test. (command "Find sub-networks enriched with selected entities" with option "Find expression targets" in Pathway Studio).

The most significant sub-network, with p-value 2.6E-06, has a transcriptional factor REST (RE1-silencing transcription factor) as a seed. This gene encodes a transcriptional repressor which represses neuronal genes in non-neuronal tissues. The down-regulation of REST/NRSF targets in spinal cord tissue of fALS patients and transgenic mice raises the possibility that this factor is erroneously activated in SOD1-related fALS, leading to the disruption of calcium homeostasis in motor neurons. Aberrant activation of REST/NRSF has been linked to Huntington's disease [37,38], through a mechanism in which mutations in huntingtin protein abolish its ability to sequester REST/NRSF from the nucleus.

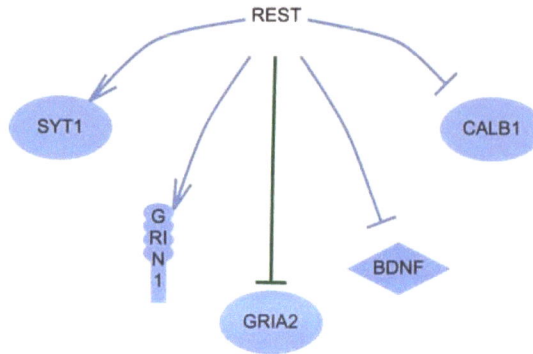

Figure 3: An Activated calcium-related sub-network regulated by REST transcriptional repressor found by SNEA.

SUMMARY

Using microarray data analysis from tissue of fALS patients and mutant SOD1 mice we found that differential expression of a number of genes could lead to increased calcium concentration in motor neurons (Table **1**). Among them are calcium channels and calcium ATP-ases, as well as regulatory and signaling proteins. Moreover, we found that in familial ALS the REST transcriptional factor may be activated, leading to transcriptional changes related to calcium homeostasis.

Table 1: Genes involved in calcium homeostasis and differentially expressed in ALS

Gene	Description
ANXA7	annexin A7
ARHGDIB	Rho GDP dissociation inhibitor (GDI) beta
ATP2A3	ATPase, Ca^{++} transporting, ubiquitous
ATP2B2	ATPase, Ca^{++} transporting, plasma membrane 2
BDNF	brain-derived neurotrophic factor
C5AR1	complement component 5a receptor 1
CACNA1C	calcium channel, voltage-dependent, L type, alpha 1C subunit
CACNB3	calcium channel, voltage-dependent, beta 3 subunit
CALB1	calbindin 1, 28kDa
GRIA2	glutamate receptor, ionotropic, AMPA 2
GRIN1	glutamate receptor, ionotropic, N-methyl D-aspartate 1
GRIN2A	glutamate receptor, ionotropic, N-methyl D-aspartate 2A
GRIN2C	glutamate receptor, ionotropic, N-methyl D-aspartate 2C

Table 1: cont....

GRK6	G protein-coupled receptor kinase 6
IGF1	insulin-like growth factor 1 (somatomedin C)
ITPR2	inositol 1,4,5-triphosphate receptor, type 2
PTAFR	platelet-activating factor receptor
RGN	regucalcin (senescence marker protein-30)
RGS7	regulator of G-protein signaling 7
RRAS	related RAS viral (r-ras) oncogene homolog

MATERIALS AND METHODS

We obtained microarray experiments from the GEO database [http://www.ncbi.nlm.nih.gov/gds] for human familial ALS samples (GDS412) and samples from fALS model SOD1-mutant mice (GSE4390 and GSE16362). For GDS412 data, the differential expression values "fALS *vs.* control" and "sALS *vs.* control"(for reference only) were calculated in Pathway Studio [39,40] using a two-tailed T-test. Because there are only two samples available for fALS, more elaborated statistical techniques for the ranking of differential expression genes were not possible.

Two experiments with samples from transgenic mice expressing SOD1 protein with the G93A mutation found in human ALS (GSE16362 and GSE4390 from the GEO database) were aggregated and several methods of differentially expressed (G93A mutant *vs.* WT mice) genes ranking were applied. First, all genes with more than 30% of missing values were removed; otherwise, missing data have been replaced using the K-nearest neighbor method with k=10 [41]. All values were log-transformed, normalized using median normalization and scaled. Then, the following gene ranking methods have been applied for each dataset separately: log-ratio, T-test, Wilcoxon test, limmaT-test [42], SAM test [43], and shrink T-test [44]. Finally, we have combined the results from different experiments to generate the single "differential" rank for each gene. We used Fisher's method to combine p-values of the same type:

$$X_{2k}^2 = -2 \sum_{i=1}^{k} \log_e (p_i).$$

,

Values of different statistics were averaged for each gene. The final combined gene ranking R was calculated as mean of the ranks from all methods. Each gene was also assigned a single differential log ratio value calculated as an average differential log-ratio from all gene expression datasets. The resulting two experiments (human GDS412 and mouse aggregated) were analyzed in Pathway Studio software.

We used ResNet 7.0 [40] and ChemEffect 2.0 [45] databases of biological relations for our network and pathway analysis, which is automatically constructed and updated using MedScan natural language processing (NLP) technology [46] from PubMed-indexed abstracts and full-text articles. MedScan Technology allows the extraction of relationships between cellular processes, diseases, functional classes, complexes, proteins, and small molecules in the ResNet database, along with corresponding citations. We searched ResNet for Cell Process-centered sub-networks that are enriched in genes differentially expressed in ALS using Pathway Studio Enterprise edition and the implemented Sub-network Enrichment Analysis (SNEA) algorithm with Mann-Whitney or Fisher exact tests [47]. Cellular Processes are defined in ResNet as Gene Ontology terms or terms from the proprietary Ariadne Ontology.

CONFLICT OF INTEREST

Authors do not have any conflicts of interests with respect to chapter content

ACKNOWLEDGEMENT

This work was partially supported by the grant from the US National Institutes of Health to N.V.D. (GM080742).

REFERENCES

[1]   Redler RL, Dokholyan NV. The Complex Molecular Biology of Amyotrophic Lateral Sclerosis (ALS). Prog Mol Biol Transl Sci 2012;107:215–62.
[2]   Banci L, Bertini I, Boca M, Girotto S, Martinelli M, Valentine JS, Vieru M. SOD1 and amyotrophic lateral sclerosis: mutations and oligomerization. PLoS ONE 2008;3:e1677.
[3]   Proctor EA, Ding F, Dokholyan NV. Structural and thermodynamic effects of post-translational modifications in mutant and wild type Cu, Zn superoxide dismutase. J. Mol. Biol. 2011;408:555–67.

[4]   Alexianu ME, Kozovska M, Appel SH. Immune reactivity in a mouse model of familial ALS correlates with disease progression. Neurology 2001;57:1282–9.

[5]   Henkel JS, Beers DR, Zhao W, Appel SH. Microglia in ALS: the good, the bad, and the resting. J Neuroimmune Pharmacol 2009;4:389–98.

[6]   Damiano M, Starkov AA, Petri S, Kipiani K, Kiaei M, Mattiazzi M, Flint Beal M, Manfredi G. Neural mitochondrial Ca^{2+} capacity impairment precedes the onset of motor symptoms in G93A Cu/Zn-superoxide dismutase mutant mice. J. Neurochem. 2006;96:1349–61.

[7]   Kawamata H, Manfredi G. Mitochondrial dysfunction and intracellular calcium dysregulation in ALS. Mech. Ageing Dev. 2010;131:517–26.

[8]   Jaiswal MK, Zech W-D, Goos M, Leutbecher C, Ferri A, Zippelius A, Carrì MT, Nau R, Keller BU. Impairment of mitochondrial calcium handling in a mtSOD1 cell culture model of motoneuron disease. BMC Neurosci 2009;10:64.

[9]   Alexianu ME, Ho BK, Mohamed AH, La Bella V, Smith RG, Appel SH. The role of calcium-binding proteins in selective motoneuron vulnerability in amyotrophic lateral sclerosis. Ann. Neurol. 1994;36:846–58.

[10]   Ho BK, Alexianu ME, Colom LV, Mohamed AH, Serrano F, Appel SH. Expression of calbindin-D28K in motoneuron hybrid cells after retroviral infection with calbindin-D28K cDNA prevents amyotrophic lateral sclerosis IgG-mediated cytotoxicity. Proc. Natl. Acad. Sci. U.S.A. 1996;93:6796–801.

[11]   Shaw PJ, Eggett CJ. Molecular factors underlying selective vulnerability of motor neurons to neurodegeneration in amyotrophic lateral sclerosis. J. Neurol. 2000;247 Suppl 1:I17–27.

[12]   Pottorf WJ, De Leon DD, Hessinger DA, Buchholz JN. Function of SERCA mediated calcium uptake and expression of SERCA3 in cerebral cortex from young and old rats. Brain Res. 2001;914:57–65.

[13]   van Es MA, Van Vught PW, Blauw HM, Franke L, Saris CG, Andersen PM, Van Den Bosch L, de Jong SW, van 't Slot R, Birve A, Lemmens R, de Jong V, Baas F, Schelhaas HJ, Sleegers K, Van Broeckhoven C, Wokke JHJ, Wijmenga C, Robberecht W, Veldink JH, Ophoff RA, van den Berg LH. ITPR2 as a susceptibility gene in sporadic amyotrophic lateral sclerosis: a genome-wide association study. Lancet Neurol 2007;6:869–77.

[14]   Van Den Bosch L, Verhoeven K, De Smedt H, Wuytack F, Missiaen L, Robberecht W. Calcium handling proteins in isolated spinal motoneurons. Life Sci. 1999;65:1597–606.

[15]   Dayanithi G, Mechaly I, Viero C, Aptel H, Alphandery S, Puech S, Bancel F, Valmier J. Intracellular Ca^{2+} regulation in rat motoneurons during development. Cell Calcium 2006;39:237–46.

[16]   Smith RG, Alexianu ME, Crawford G, Nyormoi O, Stefani E, Appel SH. Cytotoxicity of immunoglobulins from amyotrophic lateral sclerosis patients on a hybrid motoneuron cell line. Proc. Natl. Acad. Sci. U.S.A. 1994;91:3393–7.

[17]   Kimura F, Smith RG, Delbono O, Nyormoi O, Schneider T, Nastainczyk W, Hofmann F, Stefani E, Appel SH. Amyotrophic lateral sclerosis patient antibodies label Ca^{2+} channel alpha 1 subunit. Ann. Neurol. 1994;35:164–71.

[18]   Appel SH, Smith RG, Engelhardt JI, Stefani E. Evidence for autoimmunity in amyotrophic lateral sclerosis. J. Neurol. Sci. 1993;118:169–74.

[19]   Canton T, Pratt J, Stutzmann JM, Imperato A, Boireau A. Glutamate uptake is decreased tardively in the spinal cord of FALS mice. Neuroreport 1998;9:775–8.

[20]   Tateno M, Sadakata H, Tanaka M, Itohara S, Shin R-M, Miura M, Masuda M, Aosaki T, Urushitani M, Misawa H, Takahashi R. Calcium-permeable AMPA receptors promote

misfolding of mutant SOD1 protein and development of amyotrophic lateral sclerosis in a transgenic mouse model. Hum. Mol. Genet. 2004;13:2183–96.

[21] Rothstein JD, Tsai G, Kuncl RW, Clawson L, Cornblath DR, Drachman DB, Pestronk A, Stauch BL, Coyle JT. Abnormal excitatory amino acid metabolism in amyotrophic lateral sclerosis. Ann. Neurol. 1990;28:18–25.

[22] Samarasinghe S, Virgo L, de Belleroche J. Distribution of the N-methyl-D-aspartate glutamate receptor subunit NR2A in control and amyotrophic lateral sclerosis spinal cord. Brain Res. 1996;727:233–7.

[23] Watson WD, Srivastava M, Leighton X, Glasman M, Faraday M, Fossam LH, Pollard HB, Verma A. Annexin 7 mobilizes calcium from endoplasmic reticulum stores in brain. Biochim. Biophys. Acta 2004;1742:151–60.

[24] Tobisawa M, Tsurusaki Y, Yamaguchi M. Decrease in regucalcin level and enhancement of protein tyrosine phosphatase activity in rat brain microsomes with increasing age. Int. J. Mol. Med. 2003;12:577–80.

[25] Yamaguchi M, Hanahisa Y, Murata T. Expression of calcium-binding protein regucalcin and microsomal Ca^{2+}-ATPase regulation in rat brain: attenuation with increasing age. Mol. Cell. Biochem. 1999;200:43–9.

[26] Magrané J, Rosen KM, Smith RC, Walsh K, Gouras GK, Querfurth HW. Intraneuronal beta-amyloid expression downregulates the Akt survival pathway and blunts the stress response. J. Neurosci. 2005;25:10960–9.

[27] Vincent AM, Mobley BC, Hiller A, Feldman EL. IGF-I prevents glutamate-induced motor neuron programmed cell death. Neurobiol. Dis. 2004;16:407–16.

[28] Kleppisch T, Klinz FJ, Hescheler J. Insulin-like growth factor I modulates voltage-dependent Ca^{2+} channels in neuronal cells. Brain Res. 1992;591:283–8.

[29] Dolcet X, Egea J, Soler RM, Martin-Zanca D, Comella JX. Activation of phosphatidylinositol 3-kinase, but not extracellular-regulated kinases, is necessary to mediate brain-derived neurotrophic factor-induced motoneuron survival. J. Neurochem. 1999;73:521–31.

[30] Caldeira MV, Melo CV, Pereira DB, Carvalho RF, Carvalho AL, Duarte CB. BDNF regulates the expression and traffic of NMDA receptors in cultured hippocampal neurons. Mol. Cell. Neurosci. 2007;35:208–19.

[31] Feuerstein G, Yue TL, Lysko PG. Platelet-activating factor. A putative mediator in central nervous system injury? Stroke 1990;21:III90–94.

[32] Bennett SA, Chen J, Pappas BA, Roberts DC, Tenniswood M. Platelet activating factor receptor expression is associated with neuronal apoptosis in an *in vivo* model of excitotoxicity. Cell Death Differ. 1998;5:867–75.

[33] Witke W, Li W, Kwiatkowski DJ, Southwick FS. Comparisons of CapG and gelsolin-null macrophages: demonstration of a unique role for CapG in receptor-mediated ruffling, phagocytosis, and vesicle rocketing. J. Cell Biol. 2001;154:775–84.

[34] Luo J, Busillo JM, Benovic JL. M3 muscarinic acetylcholine receptor-mediated signaling is regulated by distinct mechanisms. Mol. Pharmacol. 2008;74:338–47.

[35] Shuey DJ, Betty M, Jones PG, Khawaja XZ, Cockett MI. RGS7 attenuates signal transduction through the G(alpha q) family of heterotrimeric G proteins in mammalian cells. J. Neurochem. 1998;70:1964–72.

[36]   Saitoh O, Kubo Y, Odagiri M, Ichikawa M, Yamagata K, Sekine T. RGS7 and RGS8 differentially accelerate G protein-mediated modulation of K+ currents. J. Biol. Chem. 1999;274:9899–904.

[37]   Zuccato C, Tartari M, Crotti A, Goffredo D, Valenza M, Conti L, Cataudella T, Leavitt BR, Hayden MR, Timmusk T, Rigamonti D, Cattaneo E. Huntingtin interacts with REST/NRSF to modulate the transcription of NRSE-controlled neuronal genes. Nat. Genet. 2003;35:76–83.

[38]   Zuccato C, Belyaev N, Conforti P, Ooi L, Tartari M, Papadimou E, MacDonald M, Fossale E, Zeitlin S, Buckley N, Cattaneo E. Widespread disruption of repressor element-1 silencing transcription factor/neuron-restrictive silencer factor occupancy at its target genes in Huntington's disease. J. Neurosci. 2007;27:6972–83.

[39]   Yuryev A, Mulyukov Z, Kotelnikova E, Maslov S, Egorov S, Nikitin A, Daraselia N, Mazo I. Automatic pathway building in biological association networks. BMC Bioinformatics 2006;7:171.

[40]   Nikitin A, Egorov S, Daraselia N, Mazo I. Pathway studio--the analysis and navigation of molecular networks. Bioinformatics 2003;19:2155–7.

[41]   Troyanskaya O, Cantor M, Sherlock G, Brown P, Hastie T, Tibshirani R, Botstein D, Altman RB. Missing value estimation methods for DNA microarrays. Bioinformatics 2001;17:520–5.

[42]   Smyth GK. Linear models and empirical bayes methods for assessing differential expression in microarray experiments. Stat Appl Genet Mol Biol 2004;3:Article3.

[43]   Tusher VG, Tibshirani R, Chu G. Significance analysis of microarrays applied to the ionizing radiation response. Proc. Natl. Acad. Sci. U.S.A 2001;98:5116–21.

[44]   Opgen-Rhein R, Strimmer K. Accurate ranking of differentially expressed genes by a distribution-free shrinkage approach. Stat Appl Genet Mol Biol 2007;6:Article9.

[45]   Kotelnikova E, Yuryev A, Mazo I, Daraselia N. Computational approaches for drug repositioning and combination therapy design. J Bioinform Comput Biol 2010;8:593–606.

[46]   Daraselia N, Yuryev A, Egorov S, Novichkova S, Nikitin A, Mazo I. Extracting human protein interactions from MEDLINE using a full-sentence parser. Bioinformatics 2004;20:604–11.

[47]   Sivachenko AY, Yuryev A, Daraselia N, Mazo I. Molecular networks in microarray analysis. J Bioinform Comput Biol 2007;5:429–56.

SUPPLEMENTARY DATA

Supplementary Table 1: Cell processes found to be enriched with genes differentially expressed in fALS analysis in ResNet 8.

Gene Set Seed	Total # of genes	# of Measured genes	Measured genes
cell recognition	82	62	SELL, MBP, HLA-F, CSF2, CD80, CD8A, CD14, CD1D, CTLA4, CD1A, SLC4A1, CCL21, HLA-E, LST1, C TLR4, PROS1, CD44, CD48, FN1, ITGAM, IL2, ACAN, CD4, KLRK1, ICAM1, ITGB1, VEGFA, HLA-DQA ITGB2, PITX2, CD1C, CDH1, IL15, CD7, CD99, TYR, TNFSF4, ANXA1, PLG, CCR7, TRA@, LTF, CDH13 PLAUR, IRF8, HSPA1A, CD1B, IRF1, HLA-G, B2M, NCAM1, CYCS, TP53, PTPRC, HLA-A, CD86
monocyte differentiation	73	59	IL6, IL8, VIP, IL1B, PF4, CSF2, TNF, ARF6, CD14, ADORA2A, IRF7, MAPK1, CD1D, INHBA, CEBPA, TC TGM2, FN1, ACAT1, CTSS, IL2, IFIT3, PPARA, IL19, VEGFA, FAS, PRKCD, IL17A, RARA, LIF, IL32, M KLF4, LGALS3, RXRA, PTK2B, SELP, CDKN1A, CCL2, VDR, IFNG, NEU1, CCR5, REL, PPARG, MMP9, NCAM1, CSF1, IL4, PTPRC, CDK9, CD40
superoxide release	38	36	NOS2A, IL8, MAPK8, AHSG, CSF2, AKT1, PRKCA, TNF, CD14, MAPK1, CRP, PLAU, CSF3, PRKCB, CC PLAUR, PDE1A, MAPK14, AGER, FPR1, ITGAM, IL2, ALB, ITGA2B, NOS3, FCGR1A, F2, TNFRSF1A, A CYCS, ACE, PRTN3
donor preference	9	7	ZNF143, HLA-DPB1, HLA-DRB1, HLA-DQB1, HLA-C, FOXO1, HLA-A
rosetting	27	25	IL6, LGALS3, G6PD, CD99, CD8A, CR2, SELP, CD33, CFD, NEU1, CD2, MPO, CD44, CD48, FN1, ITGAM FCGR1A, CD58, F2, FCGR2A, FCGR2B, CR1
anoikis	153	120	F7, PAK1, PRKCA, TNF, CAV1, RAC1, SERPINF1, MAPK1, RHOB, CTSG, PTGS2, PARP1, TDGF1, DAP ITGB3, ERBB2, NTRK2, CASP8, EGFR, BIRC3, MAPK14, BCL2L1, ITGAV, BDNF, PLAT, KRAS, FAS, C NTRK3, PDGFRB, HRAS, TNFRSF11B, TGFA, CDKN2A, CDH1, SMAD4, CSPG4, MAPK8, YWHAZ, LG, EGR1, ELA2, SRC, PLG, PLAU, HGF, IGF1, FASLG, TNFSF10, CDKN1A, CD2, RAF1, BAK1, KLF5, CAS TP53, SERPINE1, INSR, AREG, LPAR1, RHOA, IGF1R, MCL1, SLC3A2, RALGDS, CTGF, TPM2, ILK, M ITGA4, TGFB1, CCND1, TIMP1, MAPK11, VSNL1, FN1, CEACAM5, BIRC5, ITGB1, TIAM1, PTEN, CDC NPPA, RBP1, CASP9, CTNNB1, IGF2, BCL2, AKT1, PTK2B, BSG, ARHGDIB, ESR2, CFLAR, FADD, CXC MTA1, MAP2K4, GRN, NGF, PXN, PTPN11, SHC1, BRAF, DAP3
calcium-mediated signaling	55	48	PKD1, BCR, LEP, CALB1, CD38, P2RX7, TNF, ACCN2, F2RL1, ADCYAP1, LEPR, AGT, NTRK2, UCHL1 FCGR1A, BDNF, F2, CD4, ITGB1, ITPR2, CCK, ITPR1, AGTR1, RIT2, ATP2A1, GRM5, PPP3R1, NTS, AN HINT1, PLAU, GCG, IGF1, PLAUR, PTH, FGFR2, SLC8A1, APOE, PLCG1, NGF, HRH1, TPT1, CR1
complement activation	96	73	C1R, PTGIR, HMOX1, MCM5, MBL2, IL8, SELL, PRNP, LHCGR, IL1B, BGN, BPI, TNF, ICAM3, AMBP, FMOD, CRP, COL17A1, CD52, CFD, CFP, AGT, COL2A1, C5, PROS1, MMP14, SELPLG, BCL2L1, FN1, C ICAM1, APP, CD97, C5AR1, C3, ITGB2, LYZ, COL4A3, CD55, BCL2, ELA2, PTX3, C1QBP, PLG, SELP, C F12, SERPINC1, LTF, CFI, PHB, CD46, SERPING1, DEFA1, FCN2, A2M, C4BPA, APOE, MASP1, B2M, FC CR1
osteoclast	158	117	IL6, TSHR, CYR61, CSF2, TNF, CNR2, NFATC1, SPP1, TRAF6, WNT10B, RAC1, MAPK1, INHBA, PTGS

| differentiation | | | LRRC17, CASP3, CALCA, EGFR, KIR2DL3, TLR4, EPHB4, MAPK14, FLT1, OSM, FAS, FGF2, PTPRO, TCF
FOSL1, TGFA, EPHA2, IL11, PPP3CA, NOS2A, MAPK8, CA2, MYC, EGR1, CTF1, SRC, BMP2, GRIN1, PGF
CDKN1A, LGMN, BTG2, IFNG, SPARC, AKT2, IL3, CDH11, RUNX2, DMP1, CDK6, CXCL11, PTGER4, TG
FRAP1, IFNB1, CASR, TGFB2, IL8, LEP, FHL2, PTHR1, IL1B, BGN, MSR1, PTHLH, MDK, IL6R, FGF8, TG
CCL5, GAS6, SPI1, CD4, CSF1R, VEGFA, PTEN, IL17A, KL, HIF1A, LIF, IBSP, CTNNB1, IL15, IGF2, JUNB
FOS, MAPKAPK2, CCL2, FLT4, MAPK3, CXCL12, PPARG, PTH, CCL4, TNFRSF1A, PLCG1, IL1A, MITF, T |
| neutrophil
homeostasis | 5 | 5 | ST6GAL1, CXCL12, IL17A, CXCR4, CSF3 |

Supplementary Table 2: Cell processes found to be enriched with genes differentially expressed in SOD network enrichment analysis in ResNet 8.

Gene Set Seed	Total # of genes	# of Measured genes	Measured genes
microglial activation	164	135	SOD1, IL6, IL1R1, IL18, TNF, CNR2, CD14, VTN, SPP1, FYN, MAPK1, F2RL3, DDR1, PTGS2, IL12A, CLU EGFR, MAPK14, EDN1, CD44, AGER, SNCA, DDC, BDNF, F2, GHRL, PLAT, FAS, EPHA4, APP, CXCR4, NOS2A, MAPK8, SERPINI1, SRC, CXCL2, SOCS1, WAS, PLG, IDE, GRIN1, WNT1, FGF1, IGF1, ANXA2, TSPO, C1QA, F10, SLC8A1, RAF1, APOE, STAT3, FCER1G, GABBR1, MMP3, TP53, PTPRC, CREB1, CD4 IL1B, CD38, P2RX7, IFNGR1, MIF, CD8A, MSR1, ADORA2A, SERPINA3, CAT, TGFB1, ADCYAP1, IL1RI GAS6, FN1, ITGAM, CTSS, TAC1, IL4R, TRADD, PPARA, CD4, CSF1R, CHRNA7, SOCS3, S100B, RELA, MYD88, CD40LG, ESR1, CASP9, CEBPB, IL13, STAT1, RGS10, KCNN4, NTF3, ESR2, FPR2, CD200, MAP PPARG, MMP9, GFAP, CTSB, PTPN6, CD36, TNFRSF1A, IL10, IL1A, B2M, CSF1, APOD, IL4, VIM, CD86
hemopoie sis	57	48	IKZF1, LMO2, CXCL9, IL6, LEP, IL1B, CSF2, TNF, TRAF6, VCAM1, CD1D, CSF3, KITLG, GJA1, TGFB1, TAL1, CASP8, JAK2, CD44, IL2, TAC1, IKZF2, CD4, GHRL, VEGFA, IL17A, PECAM1, CXCR4, LIF, TGFA BMP4, IL13, MZF1, HGF, IGF1, TNFSF10, FADD, EPO, CXCL12, KIT, IFNA1, TIE1, IL1A, CSF1
anaphylax is	58	37	TRH, HMOX1, IL13RA1, NOS2A, JAK3, IL18, LYZ, BCL10, WIPF1, CSF2, IL13, SPP1, CD63, TGFB1, RCA STAT6, RASGRP1, TAC1, LALBA, ALB, IL4R, FCGR1A, NOS3, F2, FCGR2A, IL10, PECAM1, FCGR2B, F IL9, PTAFR, FCER1A
mast cell activation	103	64	MET, CMA1, SYK, BCL10, WIPF1, PIK3CD, CSF2, SPHK2, CR2, MAPK1, CAT, F2RL3, PRKCB, KITLG, F LAT, INPP5D, CCL5, EDN1, MAPK14, TUBG1, THY1, LAT2, TAC1, ALB, FCGR2A, ICAM1, ITGB1, KRA: PECAM1, FCGR2B, C5AR1, CCKBR, FCER1A, GATA2, C3AR1, PTPRE, S100A8, TNFSF4, EGR1, PLD2, C MAPK3, KIT, HCK, Swap70, IL5, SH3BP2, IL10, STAT3, SIRPA, PLD1, FCER1G, NGF, PTPN11, IL4, IL9, E
degranula tion	292	211	SYT2,SYK,BCL10,PTGER1,CSF2,PRKCA,TNF,VTN,LCP2,RAC1,FYN,MAPK1,CRP,VCAM1,CD63,CFTR,(L7,INPP5D,LAT,C5,EDN1,ORAI1,ALB,F2,KLRK1,NDRG1,PLAT,PRKCD,GPR44,APP,C3,PTGER2,PTAFR, USP1,ANXA1,VAMP7,ELA2,SLC2A4,TYROBP,PLG,PLAU,IFNG,CCL13,FLOT1,CD2,PTPRJ,LTB4R,VAM 11,LAMP1,ADM,PTPRC,SFTPD,CD40,CMA1,VIP,SELL,SP1,CSF3R,IAPP,SNAP23,IL1B,TNFSF8,RNASE3, AT,CSF3,PRKCB,GZMB,ADK,SLC9A1,KITLG,TGFB1,CFD,ADCYAP1,CCL5,LAT2,FN1,FPR1,ITGAM,IL2 ,ICAM1,ITGB1,PTEN,STX4,LIF,EDG5,CX3CL1,IL15,MS4A2,CCR3,PTK2B,SYT1,BSG,RASA4,FOS,F12,T? CD200,GNAQ,MAPK3,HOXA10,CXCL3,KIT,PLAUR,MMP9,PTPN6,CCL4,KLRB1,IL10,CSK,GRB2,IL4,SL K1,PIK3CD,SH3KBP1,SPP1,CTSG,F2RL1,ANG,PNOC,MAPK14,KRAS,FCGR2B,CANT1,GATA2,ITGB2,CI K2,LGALS3,GAST,SRC,WAS,PRF1,CD81,APOA1,TRPV2,FASLG,PLD1,FCER1G,VWF,SAG,CAMP,BTK,A WIPF1,PF4,CD38,PLCG2,ADORA2A,NMU,SPHK2,CR2,RAB3A,GAB2,ITGA4,Art2b,PIK3CA,TACR1,CD48

			,C5AR1,SRGN,DGKZ,FCER1A,INS,NPPA,JUNB,GATA1,RB1,RAB3D,PLD2,MYLK,PPARG,HCK,Swap A1,PLCG1,IL1A,NGF,CPLX2,SERPINB9,PRTN3
calcium mobilizati on	445	309	PKD1,F7,IL6,SYK,GRPR,PTGER1,PROC,CSF2,PRKCA,DRD2,TNF,LCP2,HTR1A,MAPK1,FYN,CRP,OX H1,CCL7,INPP5D,LAT,LEPR,AGT,EGFR,HBEGF,EDN1,BAX,BCL2L1,P2RY2,CD44,ALB,CALR,BDNF HRL,PLAT,FAS,P2RY1,PRKCD,GPR44,PECAM1,FGF2,CXCR4,PRL,PTAFR,GIT1,F2R,SAA1,IL8RB,LY 247,KCNJ6,EGF,KCNJ3,GNAI3,KCNJ5,ELA2,TYROBP,PLG,SELP,PLAU,HGF,CCR7,PGF,BIK,OPRD1,N FLOT1,CD2,PTPRJ,BDKRB2,CD27,LTB4R,CXCR7,SLC8A1,RYR1,CCL11,CD5,ADM,AVP,PTPRC,RYR L9,GNA15,ZAP70,VIP,SELL,MCAM,IL16,IL1B,TXK,EEF2K,CD8A,PDCD1,AMBP,CD9,TAC3,RGS7,ST B,MED22,KITLG,GNB5,TGFB1,ADCYAP1,CCL5,GNAI1,OPRM1,LAT2,SPN,ITGAM,FPR1,SH2D2A,IL A,CD4,ICAM1,CCL25,VEGFA,CHRNA7,PTEN,CCK,HTR2C,EDG5,CX3CL1,CCKBR,SCYE1,MYD88,G K2B,SELE,BSG,ARHGDIB,PPIB,GALR2,CD3E,FPR2,PTGFR,GNAQ,NPY,EPO,KDR,OPRS1,KLRA1,CX H,MMP9,ADRA1B,PTPN6,CCL4,CXCL14,PTPN11,CSK,EDNRB,GRB2,IL4,CSNK1A1,CD72,POMC,IKZ R1,CD79A,PIK3CD,PLCD1,ADCYAP1R1,SH3KBP1,BMX,PRKCQ,F2RL3,CTSG,CRHR1,F2RL1,CCR9,C R,BLNK,TBXA2R,PRKD1,FCGR2B,Klra4,ITPR1,SSTR5,S100A9,DPP4,AVPR2,ITGAL,RGS3,ITGB2,DN VASP,GAST,EDN3,SRC,CXCL2,CXCR3,CCL27,CD81,CEL,TRPV2,FASLG,KCNB1,KLRG1,CCKAR,AC PIP5K3,PLA2G1B,FCER1G,HOMER1,VWF,PTPN1,TP53,CAMP,ADRBK1,FGFR1,BTK,LPAR1,GRAP2, IK3C2A,CD38,PPL,PLCG2,PTHLH,Ccl6,ADORA2A,VAV2,NMU,MS4A1,GRK6,RIC8A,GAB2,CCL17,T 4,TAC1,SST,CCL1,RGS8,CD22,BRS3,C5AR1,CCR4,GHRH,OPRK1,FCER1A,INS,TRH,ADA,ADRA2A,C MK2A,GRM8,CNR1,Klra7,GCG,CFLAR,NMB,MYLK,ADORA1,PLCG1,PARP2,NGF,TNFSF11,CD180,P
B-cell activation	170	125	IL6,SYK,TSHR,CDKN2C,CD79A,PIK3CD,TNF,CNR2,LTA,OGT,NFATC1,SPP1,CD70,VCAM1,PRKCQ 4,CD44,SERPINA1,CD53,BLNK,FAS,PRKCD,PECAM1,FCGR2B,C3,HRAS,LPL,POU2F2,MAPK8,KLF4 FOXJ1,SRC,HGF,CD81,CCR7,FASLG,IL2RA,TNFRSF17,TRAF3,ENPEP,IFNG,REL,NEU1,FCER2,VAV 27,IL5,IFNA1,BHLHB2,RYR1,PIK3R1,CD5,PTPRC,IL9,CREB1,BTK,CD40,TNFRSF4,CD38,CD80,PLCG 6GAL1,CD59,CR2,PTPN12,SEMA4D,IL6R,PRKCB,NFKB1,STRA13,SPI1,POU2AF1,NOTCH1,CD48,IL2 A,ICAM1,TLR7,INS,MYD88,CD40LG,HSPD1,IL15,TNFRSF21,BCL2,IL13,STAT1,PAX5,SPIB,TNFRSF8 F3,STAT6,Swap70,PTPN6,TLR6,PLCG1,IL10,NGF,MITF,PRDM1,PTPN11,CD180,IL4,UNG,AIRE,CD72,
Leukotrie ne synthesis	37	31	AFP,JAK3,IL6,ALOX12,SYK,LEP,CSF2,EGF,TNF,MAPK1,ALOX5AP,GPX4,CSF3,PTGS2,PTGS1,TGFB A,AGT,SPARC,LTC4S,EDN1,MAPK14,IL5,TAC1,SST,PLA2G5,CCL11,COTL1,FCER1A
T-cell proliferati on	431	317	SDC4,LMO2,IL6,BCL10,CDKN2C,ABL1,PROC,CSF2,SEMA7A,GSK3B,TNF,NFATC1,MAPK1,SERPIN D70,PTGS2,INHBA,CD63,ITGAX,CD19,GNRH1,FLT3LG,CASP3,AGT,COL2A1,CASP8,TAL1,TCL1A,N CD44,ALB,F2,IKZF2,KLRK1,FAS,CCNE1,APP,EIF2AK2,PRL,HLA- DQA1,Bim,PTGER2,LY75,HSPA8,LYZ,CYP21A2,TNFRSF1B,PDE7A,ANXA1,FGL2,SOCS1,CST3,HGF PRD1,IFNG,FCER2,CDKN1B,CD46,CD2,PTPRJ,CD27,LTB4R,SHH,CD83,MAP3K8,BAK1,STAT3,CD5,6 8,FRAP1,SFTPD,IL15RA,PTPRC,IL9,TGFB2,ARG1,CD40,KRT7,RASSF5,HMOX1,CXCL9,MMP2,MBL2 2,IL16,TNFSF8,IL1B,PRNP,CD80,IFNGR1,CD8A,MIF,PDCD1,CD9,CD1D,CAT,CSF3,HSP90B1,TGFB1, GDS,SELPLG,SPI1,THY1,FN1,SPN,ITGAM,CTSS,IL2,MYB,BIRC5,FCGR1A,TNC,CD4,PPARA,ITGB1,N FA,PTEN,SOCS3,IL17A,MUC1,CD6,KCNA3,RELA,LIF,CX3CL1,RNASE2,DLL1,ADAM10,MYD88,CD4 CEBPB,NRP1,CD47,IL13,TYK2,IL4I1,BSG,KCNN4,FOS,LTF,TNFRSF8,CD3E,EFNB1,CD200,NPY,MAF ,LCK,PTH,MMP9,CTSB,EZH2,MAP2K4,RNF128,PLA2G6,CCL4,BMI1,KLRB1,IL10,E2F1,GFI1,SIRPA,I AF,CD86,POMC,JAK3,IL18,CCR1,INDO,CIITA,ICOS,TSC22D3,CD14,ARF6,CAV1,SPP1,PRKCQ,SLC4 22,ITK,HMGA1,MAPK9,PNOC,JAK2,MAPK14,CD53,TBXA2R,CD34,MOG,FCGR2B,TNFRSF9,DPP4,R MAPK8,LGALS3,MYC,TNFSF4,FOXJ1,SRC,WAS,CD81,RPLP2,PRF1,FGF1,FASLG,IL2RA,CDKN1A,T EL,VAV1,IRAK1,SIGLEC1,GADD45A,CDK4,IFNA1,IL10RA,APOE,HLA- G,RASSF3,TRAF1,TP53,ADRB1,IL17RA,IFNB1,CD37,BTK,IGF1R,TNFRSF4,LEP,IL7,MBP,DBNL,PF4,

			A,LAG3,SEMA4D,APOH,HSP90AA1,NR4A1,PRKCH,TACR1,NOTCH1,CD48,TAC1,PIM1,SST,TRAF2,HLA DMA,CD22,TERT,LGALS1,CCR4,DOCK2,INS,PREP,ADA,TNFRSF21,CD7,BCL2,C1QBP,CFLAR,FADD,IL 6,STAT5A,A2M,CD36,TNFRSF1A,EFNA1,IL1A,RORC,TNFSF11,CSF1,PRTN3,GNAI2
monocyte differentia tion	73	62	IL6,VIP,IL1B,PF4,CSF2,TNF,ARF6,CD14,ADORA2A,MAPK1,IRF7,CD1D,INHBA,CEBPA,TGFB1,CASP3,F GM2,FN1,ACAT1,CTSS,IL2,IFIT3,PPARA,VEGFA,FAS,PRKCD,IL17A,RARA,LIF,DLL1,MYD88,IL15,PPA S3,RXRA,PTK2B,TYROBP,SELP,CDKN1A,VDR,IFNG,REL,NEU1,HOXA10,CDKN1B,PPARG,MMP9,IFNA CAM1,CSF1,IL4,PTPRC,CDK9,CD40

Supplementary Table 3: Genes differentially expressed in ALS datasets from Ca^{2+} overview pathway.

Function	Affymetrix probe ID	Gene Name	fALS/normal log-ratio	p-values for fALS/normal	sALS/normal log-ratio	p-values for sALS/normal	Mouse rank
Ca^{++} ER import, Ca-ATPase	Z69881_at	ΛTP2A3	-2.8919	3.13E-01	-2.2012	2.93E-01	188
Ca^{++} export, Ca-ATPase	X63575_s_at	ATP2B2	-1.1534	6.09E-01	2.1808	1.46E-01	1023
voltage-dependent Ca^{++} import	M92269_f_at	CACNA1C	-1.342	5.42E-01	-1.0299	4.30E-01	2042
voltage-dependent Ca^{++} import	U07139_at	CACNB3	3.8729	2.00E-03	2.5792	1.01E-01	3275
calbindin	M19878_at	CALB1	-5.2747	3.55E-03	-1.0132	4.72E-01	885
calbindin	M19878_s_at	CALB1	-1.6885	3.79E-01	0.045	9.78E-01	885
calbindin	X06661_at	CALB1	-2.4791	1.69E-03	-1.9578	3.96E-02	885
ER Ca^{++} release, ITPR	D26350_at	ITPR2	4.301	6.46E-04	3.0724	2.11E-02	1540
ER Ca^{++} release	X98330_at	RYR2	-1.3922	4.82E-01	-1.0601	4.14E-01	1059
NMDA receptor	L13266_s_at	GRIN1	-1.1151	3.92E-01	0.4115	6.55E-01	145
NMDA receptor	U09002_at	GRIN2A	-4.2209	2.88E-02	-0.4708	4.44E-01	2334
NMDA receptor	L76224_at	GRIN2C	-1.4046	1.36E-02	-0.336	2.03E-01	2234

AMPA receptor	L20814_at	GRIA2	-1.7029	1.70E-02	-0.5761	2.26E-01	213
GPCR	U76764_s_at	CD97	1.7586	1.78E-01	0.8632	3.56E-01	341
GPCR	X72304_at	CRHR1	-1.0929	5.42E-01	-1.0929	2.92E-01	232
GPCR	U18550_at	GPR3	1.6163	1.78E-01	0.2644	4.07E-01	1732
Ras	M14949_at	RRAS	6.5257	2.15E-02	4.5133	5.30E-02	10
PI3K	X80907_at	PIK3R2	-1.3231	1.42E-01	-0.1858	7.12E-01	698
PKC	D10495_at	PRKCD	5.3903	5.13E-02	5.6418	2.51E-03	2405
PKC	L01087_at	PRKCQ	1.3667	1.78E-01	0.8388	4.07E-01	1419
NADPH oxidase	M21186_at	CYBA	6.6236	7.62E-02	5.4476	1.39E-02	159
NADPH oxidase	X04011_at	CYBB	1.505	4.67E-01	1.6438	2.22E-01	2374
NADPH oxidase	M32011_at	NCF2	3.0902	2.67E-01	2.8203	8.47E-02	1750
NF-kB	S76638_at	NFKB2	3.8332	3.01E-03	1.321	2.36E-01	6450
NF-kB	L19067_at	RELA	1.1667	3.97E-01	0.577	5.05E-01	282
NF-kB	U33838_at	RELA	1.1736	4.75E-01	-2.6807	6.44E-02	282
NOS	U17327_at	NOS1	-3.5355	2.98E-02	-1.3881	1.68E-01	329
p38	L35253_s_at	MAPK14	-1.8928	1.69E-02	-0.9422	3.51E-01	238

Supplementary Table 4: Genes linked to Ca^{2+} related processes differentially expressed in various ALS-relat

Name	Description	Overview pathway Calcium in ALS	Possible function in motor neuron	Mechanism	Sentence	Medline reference
ANXA7	annexin A7		+	direct	Synexin lowers the threshold of CA^{2+} concentration required for fusion of large unilamellar vesicles of phosphatidylserine and a mixture of phosphatidylserine with phosphatidylethanolamine. synexin also increases drastically the initial rate of fusion. the initial rate of fusion increases with the quantity of synexin present in the reaction mixture., This implies that the regulation of electromechanical coupling at high systolic Ca^{2+} concentration is impaired and indicates that annexin A7 is involved in the regulation of Ca^{2+} homeostasis and/or the function of the contractile apparatus in the adult stage.	6452452:3, 11390641:10273
ARHGD IB	Rho GDP dissociation inhibitor (GDI) beta		?	indirect	Ly-GDI seems to inhibit NFAT stimulation (Fig. **6**), which correlates with its capacity to block calcium mobilization (Fig. **7**)., Intriguingly, we have recently demonstrated that the amino-terminus region of Vav1 interacts *in vitro* and *in vivo* with another potential regulator of Rho GTPases, the hematopoietic-specific guanine dissociation inhibitor (GDI) protein, Ly-GDI. 59, 81 Ly-GDI seems to inhibit NFAT stimulation, which correlates with its capacity to block calcium mobilization. 81 This activity most probably stems from its ability to inhibit members of the Rho GTPases.	12386169:10331 14592821:10202
ARHGD IB	Rho GDP dissociation inhibitor (GDI) beta		?	indirect	Ly-GDI seems to inhibit NFAT stimulation (Fig. **6**), which correlates with its capacity to block calcium mobilization (Fig. **7**)., Intriguingly, we have recently demonstrated that the amino-terminus region of Vav1 interacts *in vitro* and *in vivo* with another potential regulator of Rho GTPases, the hematopoietic-specific guanine dissociation inhibitor (GDI) protein, Ly-GDI. 59, 81 Ly-GDI seems to inhibit NFAT stimulation, which correlates with its capacity to block	12386169:10331 14592821:10202

					calcium mobilization. 81 This activity most probably stems from its ability to inhibit members of the Rho GTPases.	
ATP2A3	ATPase, Ca^{++} transporting, ubiquitous	+	+	direct	SERCA2b and SERCA3 may play an important role in the regulation of calcium homeostasis in lens epithelial cells., Recombinant SERCA3b and SERCA3f proteins similarly modulate thapsigargin-induced Ca^{2+} release and subsequent Ca^{2+} influx., Ca^{2+} uptake mediated by the sarco-endoplasmic reticulum Ca^{2+} ATPase (SERCA) significantly contributes to Ca^{2+} clearance in neurons., PERK immunoblots as well as three pharmacological tools to shift the balance in one direction or another also support the idea that glucose-regulated eIF2a phosphorylation is not primarily induced by ameliorating the folding conditions in the ER lumen, *e.g.* by promoting sarco/endoplasmic reticulum Ca^{2+} ATPase-depending calcium influx, improving the ATP/ADP ratio, or ensuring a reducing environment required for the formation of the right disulfide bonds. <more data available.>	10520214:12 15028735:10 16399701:10 17082262:10 12207029:10
ATP2B2	ATPase, Ca^{++} transporting, plasma membrane 2	+	+	direct	PMCA1 and PMCA2 may be more involved in the overall cellular Ca^{2+} homeostasis., The G293S and G283S mutations delayed the dissipation of Ca^{2+} transients induced in CHO cells by InsP 3., Its extremely high expression in the lactating mammary gland suggests that PMCA2b plays a significant role in the Ca2 homeostasis of this tissue., The abundance, cell location, high affinity for Ca2, and high constitutive activity of PMCA2bw suggest that PMCA2b is important for macro-Ca2 homeostasis in lactating tissue., The rich expression of PMCA2 within PNs and their dendritic spines (Burette *et al.,* 2003) raises the possibility that postsynaptic Ca^{2+} homeostasis and downstream signaling may	11387203:10 17234811:10 10199809:10 11029307:10 17409239:10

					be impaired in the absence of PMCA2.	
BDNF	brain-derived neurotrophic factor		+	indirect	These data suggest that BDNF modulates GABA A synaptic responses by postsynaptic activation of Trk-type receptor and subsequent Ca $^{2+}$ mobilization in the CNS., Application of BDNF to cultured hippocampal neurons induced an excitatory synaptic transmission (12), cation influx (13), generation of action potential (14), and Ca $^{2+}$ mobilization (15)., Thus, mGluRI-mediated BDNF secretion in hippocampal slices and transduced hippocampal neurons is initiated by activation of mGluRI receptors, resulting in IP 3 -mediated Ca $^{2+}$ mobilization from intracellular stores. <more data available.>	9096132:10012, 11741947:10018 11285228:10173 131101708:10149, 106101314:10197, 106101752:10197
C5AR1	complement component 5a receptor 1		-		Our data shows that SPHK1 is involved in the C5aR-mediated Ca $^{2+}$ mobilization in human macrophages., Our data show that the sphingosine kinase pathway is involved in C5a receptor-mediated Ca $^{2+}$ mobilization in human neutrophils.	15265887:10186 15302883:10258

CALB1	calbindin 1, 28kDa	+	+	direct	Furthermore, the forced expression of calbindin-D 28k increases the Ca^{2+} buffering capacity of the cells, thus partially counteracting the Ca^{2+}-mediated signaling induced by 1,25(OH) 2 D 3 and EB 1089.	12072431:101
CALB1	calbindin 1, 28kDa	+	+	direct	Furthermore, the forced expression of calbindin-D 28k increases the Ca^{2+} buffering capacity of the cells, thus partially counteracting the Ca^{2+}-mediated signaling induced by 1,25(OH) 2 D 3 and EB 1089.	12072431:101
CALB1	calbindin 1, 28kDa	+	+	direct	Furthermore, the forced expression of calbindin-D 28k increases the Ca^{2+} buffering capacity of the cells, thus partially counteracting the Ca^{2+}-mediated signaling induced by 1,25(OH) 2 D 3 and EB 1089.	12072431:101
CCL13	chemokine (C-C motif) ligand 13		-		The compounds were potent inhibitors of eotaxin- and MCP-4-induced Ca^{2+} mobilization in RBL-2H3-CCR3 cells and eosinophils., We report here that MCP-4 induces eosinophil migration and calcium mobilization and potentiates basophil histamine release.	10969084:100 9062350:1003

CCL5	chemokine (C-C motif) ligand 5		-		For example, CC chemokines, such as RANTES/CCL5 and Mip1a/CCL3, induce calcium release in US28-transfected cells (35, 36) as well as in HCMV-infected fibroblasts (37).	15546882:10050
CCR1	chemokine (C-C motif) receptor 1		-		Antagonism by U83A and U83A-Npep of CCR1-mediated induction of calcium mobilization by U83A (GSRIEGR-U83) or endogenous ligand CCL3., Interleukin-8 stimulation of CXCR1 or CXCR2 cross-phosphorylated CCR1 and cross-desensitized its ability to stimulate GTPase activity and Ca^{2+} mobilization., The results show clearly that such conditions result in partial or complete removal of the inhibitory N-terminal domain and activation of CCR1-mediated calcium mobilization and cell migration *in vitro*., BX 471 was a potent functional antagonist based on its ability to inhibit a number of CCR1-mediated effects including Ca^{2+} mobilization, increase in extracellular acidification rate, CD11b expression, and leukocyte migration. <more data available.>	16365449:10263 10734056:10010 15905581:10041 10748002:10011 17234893:10033
CD72	CD72 molecule		-		CD72 reduces Ca^{2+} mobilization induced by BCR ligation, These findings confirm that CD72 suppresses cell cycle progression and Ca^{2+} mobilization in B cells after Ag stimulation.	10640734:10122 16621999:10164

CD9	CD9 molecule		-		In contrast, the CD9-induced calcium mobilization was not altered by MßCD treatment, indicating that a substantial portion of CD9 molecules exists outside rafts and can signal independently of cholesterol-dependent lipid rafts.	15941914:10
CD97	CD97 molecule	+	-		However, attempts to reveal intracellular signaling, including Ca^{2+} influx, have been mostly unsuccessful. 2 Therefore, it remains unknown at present whether the cleavage of EMR2 and CD97 at the GPS can trigger Ca^{2+} influx or any other signaling events.	15150276:10
CSNK1 A1	casein kinase 1, alpha 1		+	indirect	We also investigated the role of casein kinase-1alpha (CK1alpha) and found that knockdown of CK1alpha increased calcium mobilization but not ERK activation.	18388243:6
CX3CL	chemokine (C-		-		As shown in Fig. 1 b, 50 n M fractalkine	11432847:10

1	X3-C motif) ligand 1				induced calcium mobilization, as determined by flow cytometry., Fractalkine mediates immediate increases in intracellular calcium mobilization in both cell types and a robust program of protein phosphorylation and enzyme activation only in microglia., Analogous to the receptor-transfected CHO cell model, differential stimulation of microglia intracellular calcium mobilization by fractalkine depended on the form present., The data shown represent mean (without inhibitor or treatment) from at least three separate experiments.(D) Effect of PT treatment on fractalkine-mediated adhesion, chemotaxis, and calcium mobilization of V28-transfected 293/EBNA-1.	10415068:10204 9724801:10223, 169100885:1011 8
CXCR4	chemokine (C-X-C motif) receptor 4		-		Activation of CXCR4 on glioma stem cells induced calcium mobilization and increased VEGF and IL-8 protein secretion., CXCR4 activation also attenuates beta-adrenergic-mediated increases in calcium mobilization and fractional shortening in cardiac myocytes., Modulation of CXCR4-mediated chemotaxis and Ca^{2+} mobilization as well as surface CXCR4 expression by cytokines., IL-8 - and MGSA-mediated cross-desensitization of CCR5 and CXCR4-mediated intracellular calcium mobilization in human monocytes., Our findings that T22 inhibited Ca^{2+} mobilization induced by PBSF/SDF-1 suggest that the structure of T22 mimics the region in PBSF/SDF-1 that is involved in the binding to CXCR4. <more data available.>	17535685:8, 17010372:7, 11668182:10081 12594210:10090 9333379:10122, 16210428:10255 17404265:10161 131101460:1009 0
DOK1	docking protein 1, 62kDa (downstream of tyrosine kinase 1)		?	indirect/downs tream	FIG. **5**. p62 dok mediates inhibition of FceRI-induced Erk1/2 activation and calcium mobilization., Overexpression of p62 dok inhibits CD2-induced Ca^{2+} mobilization We evaluated the effect of p62 dok overexpression on Ca^{2+} mobilization after CD2 and CD3 stimulation.	11970986:10252 11254695:10109

| F2R | coagulation factor II (thrombin) receptor | | ? | indirect | Additional experiments showed that PAR1-mediated calcium mobilization was partially blocked by Myr-G(13)SRI(pep) but not by the Rho kinase inhibitor Y-27632., Furthermore, the effects of alpha-thrombin and thrombin receptor activating peptides (TRAP)-6 on calcium mobilization, protein kinase C (PKC) translocation, and DNA synthesis were estimated., Thrombin and PAR1-AP induced Ca^{2+} mobilization in HUVECs and mediated release of VWF and P-selectin., Cyclic GMP-dependent protein kinase Ialpha inhibits thrombin receptor-mediated calcium mobilization in vascular smooth muscle cells. <more data available.> | 17298951:7, 9392328:3, 16332977:101 17901360:101 18040024:101 14729908:103 15545263:105 10702240:101 10837487:102 15210834:100 <more data available.> |
| FCER1 G | Fc fragment of IgE, high affinity I, receptor for; gamma polypeptide | | - | | For several functions, including calcium mobilization and cytokine production, the receptor critically depends on the associated FcR ?-chain dimer (16, 17, 18), although FcaRI unassociated with the FcR ?-chain can retain some functions (19, 20). | 16517729:100 |

FCGR1 A	Fc fragment of IgG, high affinity Ia, receptor (CD64)		-		The potential of CD33 to function as an inhibitory receptor was demonstrated by its ability to down-regulate CD64-induced calcium mobilization in U937., CD45, a transmembrane tyrosine phosphatase, when co-cross-linked with either Fc gamma RI or Fc gamma RII, could prevent Fc gamma RI and Fc gamma RII-mediated calcium mobilization and protein tyrosine phosphorylation., Fc?RI coupling to phospholipase D initiates sphingosine kinase-mediated calcium mobilization and vesicular trafficking., FcgammaRI coupling to phospholipase D initiates sphingosine kinase-mediated calcium mobilization and vesicular trafficking. <more data available.>	10887109:7, 7705403:2, 15265887:10438 15661867:10316 15879148:10504 18362174:10772 17164439:10370 17311996:10120 17709506:10147 11907092:10025 <more data available.>
FCGR2 B	Fc fragment of IgG, low affinity IIb, receptor (CD32)		-		These data indicate that Fc gamma RII stimulation induces cellular signaling events such as calcium mobilization in human basophils., CD45, a transmembrane tyrosine phosphatase, when co-cross-linked with either Fc gamma RI or Fc gamma RII, could prevent Fc gamma RI and Fc gamma RII-mediated calcium mobilization and protein tyrosine phosphorylation., Calcium mobilization can be induced in immature monocytic cells (undifferentiated U937 cells) and peripheral blood monocytes with	8527947:5, 7705403:2, 2848894:1, 1655905:3, 12077245:10299 11970986:10327 17522256:10062 15151996:10088 10903717:10276 10640734:10021 <more data available.>

					an intact IgG1 anti-FcRII antibody (CIKM5) but not by F(ab')2 fragments of this antibody. <more data available.>	
FGF9	fibroblast growth factor 9 (glia-activating factor)		?	indirect	The altered calcium signaling of achondroplasic chondrocytes was confirmed, since FGF9 treatment fails to induce calcium mobilization.	18093889:7
FGFR2	fibroblast growth factor receptor 2		+	indirect	Double electrode voltage clamp technique was used to follow precisely the calcium signalling pathway activated by FGF receptors from a normal and a carcinogenous cell environment.	9491372:0

GHR	grofwth hormone receptor		?	indirect	Specific cytoplasmic domains of the growth hormone receptor mediate Jak2 activation, metabolic actions of growth hormone, Stat activation, and calcium influx.	10990439:4
GPX1	glutathione peroxidase 1		+	indirect	We have previously shown that neurons protect themselves from H_2O_2 toxicity by clearing H_2O_2 *via* glutathione peroxidase, a process that leads to the accumulation of GSSG (Desagher *et al.,* 1996) and, subsequently, to depletion of the intracellular Ca^{2+} pool.	131101242:10012

GRIA2	glutamate receptor, ionotropic, AMPA 2	+	+	direct	Presence of the glutamate receptor 2 (GluR2) subunit prevents calcium influx through AMPA-receptor complexes; deletion of this subunit results in enhanced hippocampal long-term potentiation., Another candidate mechanism for GluR2 regulation that could modify Ca^{2+} influx is RNA editing., The relative increase in synaptic GluR2 can reduce Ca^{2+} influx by forming Ca^{2+}-impermeable AMPA receptors on their own or in combination with other subunits (Fig. **10** F)., For instance, GluR2-less neurons of the GluR2 mutant mice showed increased permeability to calcium, causing an increased calcium influx leading in turn to elevated LTP upon tetanic stimulation. <more data available.>	14573529:0, 9236229:1024 11050112:103 10531456:103 10725374:101 11826150:101 17481398:100 16893420:100 164100232:10 6, 131104727:10 5 <more data available.>
GRIN1	glutamate receptor, ionotropic, N-methyl D-aspartate 1	+	+	direct	Furthermore, NMDAR-mediated calcium influx into active spines was reduced by Abeta oligomers., The calcium influx is mediated through N-methyl-d-aspartate receptor channels, which explains the neuron specificity of the response., Neuronal nitric oxide synthase (nNOS) is broadly expressed	17360908:6, 11016955:2, 19038221:0, 16165367:4, 9631036:2, 10949586:1, 14507892:12,

					in the brain and associated with synaptic plasticity through NMDAR-mediated calcium influx., Moreover, reduction of NMDAR-mediated current and calcium influx in YAC72 MSNs to levels seen in wild-type reduced NMDAR-mediated apoptosis proportionately to wild-type levels. <more data available.>	8552233:9, 18367598:10141 15140893:10239 <more data available.>
GRK6	G protein-coupled receptor kinase 6		+	indirect	We show that knockdown of GRK2, GRK3, or GRK6, but not GRK5, significantly increased carbachol-mediated calcium mobilization.	18388243:2
GZMB	granzyme B (granzyme 2, cytotoxic T-lymphocyte-associated serine esterase		-		As demonstrated in Fig. **4**, duodenase at concentrations up to 90 nm was unable to induce Ca^{2+} mobilization.	11856353:10153

	1)					
HTR2C	5-hydroxytryptam ine (serotonin) receptor 2C		+	indirect	The serotonin 5-HT2C receptor subtype signals through the heterotrimeric G-protein Gq to activate phospholipase C (PLC) (Chang *et al.,* 2000a), leading to the intracellular accumulation of inositol trisphosphate and subsequent calcium release., Introduction Top Abstract Introduction Experimental Procedures Results Discussion References The serotonin 5-HT 2C receptor (5-HT 2C R) signals through the heterotrimeric G-protein, G q, to activate phospholipase C (Chang *et al.,* 2000), leading to the intracellular accumulation of inositol trisphosphate and subsequent calcium release.	14722258:10 10999958:10
HTR2C	5-hydroxytryptam ine (serotonin) receptor 2C		+	indirect	The serotonin 5-HT2C receptor subtype signals through the heterotrimeric G-protein Gq to activate phospholipase C (PLC) (Chang *et al.,* 2000a), leading to the intracellular accumulation of inositol trisphosphate and subsequent calcium release., Introduction Top Abstract Introduction Experimental Procedures Results Discussion References The serotonin 5-HT 2C receptor (5-HT 2C R) signals through the heterotrimeric G-protein, G q, to activate phospholipase C (Chang *et al.,* 2000), leading to the intracellular accumulation of inositol trisphosphate and subsequent calcium release.	14722258:10 10999958:10

ICAM1	intercellular adhesion molecule 1		-		Investigations using fluo-3 clearly revealed that inhibition of CD18 and ICAM1-dependent adhesive interaction with AH70 cells by the treatment with specific monoclonal antibodies significantly attenuates calcium mobilization in Kupffer cells after the coculture.	9062344:10256
ICAM1	intercellular adhesion molecule 1		-		Investigations using fluo-3 clearly revealed that inhibition of CD18 and ICAM1-dependent adhesive interaction with AH70 cells by the treatment with specific monoclonal antibodies significantly attenuates calcium mobilization in Kupffer cells after the coculture.	9062344:10256

IGF1	insulin-like growth factor 1 (somatomedin C)		+	indirect	Transfected IGF-1 gene expression in postmitotic skeletal myocytes activates calcineurin-mediated calcium signalling by inducing calcineurin transcripts and nuclear localization of calcineurin protein.	10448862:3
ITGAL	integrin, alpha L (antigen CD11A (p180), lymphocyte function-associated		-		We thus investigated the effects of L-type calcium channel blockers on CD11a/CD18- and CD16-induced calcium mobilization.	9743356:1009

	antigen 1; alpha polypeptide)					
ITGAM	integrin, alpha M (complement component 3 receptor 3 subunit)		-		The CD11b chain of CR3 was found to participate in the oxidative burst and calcium mobilization induced by B. burgdorferi.	10608767:5
ITGB2	integrin, beta 2 (complement component 3 receptor 3 and 4 subunit)		-		Furthermore, all compounds, except rossicaside B, significantly inhibited PMA- and fMLP-induced Mac-1 (a beta2 integrin) upregulation at 50 microM but not that of fMLP-induced intracellular calcium mobilization., Clustering of the sIg-CD18 chimera induces intracellular calcium mobilization., We thus investigated the effects of L-type calcium channel blockers on CD11a/CD18- and CD16-induced	16393473:5, 11559699:10095 9743356:10090, 18787642:10157 9062344:10256

			-		calcium mobilization., In our experiments, incubation of macrophages with HBP caused an immediate Ca^{2+} mobilization, which was inhibited in the presence of the ß 2 integrin blocking mAb IB4 (Fig. **6** B). <more data available.>	
ITGB2	integrin, beta 2 (complement component 3 receptor 3 and 4 subunit)		-		Furthermore, all compounds, except rossicaside B, significantly inhibited PMA- and fMLP-induced Mac-1 (a beta2 integrin) upregulation at 50 microM but not that of fMLP-induced intracellular calcium mobilization., Clustering of the sIg-CD18 chimera induces intracellular calcium mobilization., We thus investigated the effects of L-type calcium channel blockers on CD11a/CD18- and CD16-induced calcium mobilization., In our experiments, incubation of macrophages with HBP caused an immediate Ca^{2+} mobilization, which was inhibited in the presence of the ß 2 integrin blocking mAb IB4 (Fig. **6** B). <more data available.>	16393473:5, 11559699:10● 9743356:100● 18787642:10 9062344:102:
ITPR2	inositol 1,4,5-triphosphate receptor, type 2	+	+	direct	IRAK2, MAPK1 and mevalonate kinase (Graef *et al.,* 1994) are involved in Ras signalling; NRTK2, IP3R type 1, IP3R type 2 and SERCA1 are involved in calcium signalling; TRAP 150, TRUP, EMX2-like and retinoic acid receptor beta (Dejean *et al.,* 1986) are implicated in the control of gene transcription; NMP 84p and Cyclin A (Wang *et al.,* 1990) play a role in the control of cell cycle; and MCM8 and hTERT are involved in the control of DNA replication. (2) Some genes are recurrently targeted by the HBV genome integration: IP3R and hTERT, each in two liver cancers.	160104436:1● 4

KCNB1	potassium voltage-gated channel, Shab-related subfamily, member 1		+	unrelated	That Kv2.1 is expressed at high levels in many mammalian central neurons suggests that Kv2.1 modulation by muscarinic signaling, and by other pathways that lead to Ca $^{2+}$ mobilization and calcineurin activation, may provide a widespread mechanism for dynamically regulating neuronal excitability.	16407566:10036
KIT	v-kit Hardy-Zuckerman 4 feline sarcoma viral oncogene homolog		?	indirect	In Y567F, Lyn activation on SCF stimulation decreased and C-terminal Src kinase (Csk) suppressed KIT-mediated Ca $^{2+}$ influx and cell migration, suggesting that Y567-mediated Src family kinase (SFK) activation leads to Ca $^{2+}$ influx and migration.	11964302:10008

KL	klotho		?	indirect	Imura *et al.,* (2007) now report that klotho also regulates calcium homeostasis., Through these pathways, alpha-Klotho participates in the regulation of calcium homeostasis of the CSF and blood/body fluids by its actions in the choroid plexus, parathyroid glands and DCT nephrons., Klotho also regulates calcium homeostasis through other mechanisms., Yoshida *et al.,* (29) reported that klotho may regulate Ca homeostasis *via* the action of 1,25(OH) 2 D 3 in the context of abnormal 1a-hydroxylase gene expression., As to how Klotho regulates calcium homeostasis, one of the most obvious hypotheses is that Klotho acts as an enzyme because it displays significant similarity to members of ß-glycosidase family 1. \<more data available.\>	17681143:1, 18160815:2, 18606998:101 16985213:100 14701853:100 18256383:100 12119304:102 15677572:100
KLRK1	killer cell lectin-like receptor subfamily K, member 1		-		ULBP-NKG2D/DAP10 interaction triggers calcium mobilization and activation of the PI3-K/Akt, MEK/ERK, and JAK2/STAT5 signaling pathways.	11777960:103

LGALS 3	lectin, galactoside-binding, soluble, 3		-		Effects of chemokines on galectin-3-induced Ca^{2+} mobilization in monocytes.	10925302:10168
MAPK1	mitogen-activated protein kinase 1		+	indirect	We show that in the same samples, SpD activates JNK and ERK at the same concentrations that inhibit bombesin-stimulated Ca^{2+} mobilization (EC 50 for JNK and ERK activation is 4.2 and 3.2 μ M, respectively, and IC 50 for inhibition of bombesin-induced Ca^{2+} mobilization = 3.7 μ M)., Furthermore, we have dissected the signaling pathways activated by GPCRs in HMC-1 cells and made the novel observations that their ability to stimulate chemokine production depends on the level of receptor expression, the length of signaling time, and the synergistic interaction of ERK phosphorylation, sustained Ca^{2+} mobilization and NFAT activation.	11323408:10298 11120854:10031
MMP9	matrix metallopeptidase 9 (gelatinase B, 92kDa gelatinase, 92kDa type IV collagenase)		+	indirect	MMP9 blocks phospholipase C, protein kinase C, Ca^{2+} mobilization, and thromboxane A2 production leading to the inhibition of the effects of MMP2 (46, 52, 53).	18663123:10194

NMB	neuromedin B		+	indirect	Of these peptides, NMB (EC50 approximately 1-10 microM) was the most active for stimulation of calcium mobilization.	8913368:4
NPY5R	neuropeptide Y receptor Y5		+	indirect	Antagonism of NPY-induced intracellular calcium transients in human NPY Y 5 - transfected LMTK mouse fibroblasts by CGP 71683A (Fig. **2**) The antagonistic properties of CGP 71683A were assessed by measuring the ability of this compound to inhibit NPY-induced Ca $^{2+}$ transients in mouse fibroblasts.	9854049:101
NT5E	5'-nucleotidase,		-		This is in complete concordance with the	11067892:10

	ecto (CD73)				observations that engagement of CD73 induces the rapid Ca^{2+} mobilization (19) needed for calpain activation.	
PLAU	plasminogen activator, urokinase		?	indirect	The inhibitory effect of PTx on the uPA-induced Ca^{2+} signalling observed in our experiments implies a G i/o -protein involvement in this pathway.	10491182:10227
PRKCD	protein kinase C, delta	+	+	indirect	PKC-d negatively regulates mast cell calcium mobilization and degranulation., Coincidently, the novel PKC isotype PKC-d has been demonstrated to be a negative regulator of Ag-mediated calcium mobilization and degranulation under supraoptimal conditions (29).	12024011:10158 15611277:10268

PRKCQ	protein kinase C, theta	+	+	indirect	Protein kinase C? affects Ca $^{2+}$ mobilization and NFAT cell activation in primary mouse T cells., Protein kinase C ? affects Ca2 + mobilization and NFAT cell activation in primary mouse T cells., Protein kinase C ? affects Ca $^{2+}$ mobilization and NFAT cell activation in primary mouse T cells., PKCtheta is required for normal TCR-driven calcium mobilization but not for ERK activation. (A) Purified mature CD3+ T cells (purified from pooled spleen and lymph nodes) were loaded with Fura-2 and monitored for changes in intracellular-free calcium [Ca^{2+}]i. <more data available.>	16479000:104 15528385:103 15728480:105 16210616:103 16493044:103 16670283:105 16709830:104 16751397:105 17182568:104 17312138:106 <more data available.>
PROC	protein C (inactivator of coagulation factors Va and VIIIa)		?	indirect	Recently, Domotor *et al.,* 19 reported that EPCR-bound APC induced intracellular calcium mobilization from the endoplasmic reticulum in cultured human umbilical vein endothelial cells.	15166095:101
PTAFR	platelet-activating factor receptor		+	indirect	Binding to the PAF receptor has previously been shown to stimulate phosphoinositide turnover and a rise in intracellular [Ca 2 +] (Mazer *et al.,* 1992).	11514591:102

PTK2B	PTK2B protein tyrosine kinase 2 beta		+	indirect/downs tream	The results of the present study demonstrate that two consensus SH3 binding sites in the carboxyl-terminal tail of the P2Y 2 receptor bind directly to Src and are important for P2Y 2 receptor-mediated activation of several tyrosine kinases, including Src, Pyk2, and the growth factor receptors, EGFR and platelet-derived growth factor receptor (PDGFR), but not for other P2Y 2 receptor-mediated responses, including intracellular calcium mobilization or activation of MAP kinases.	14670955:10029
RAB5A	RAB5A, member RAS oncogene family		+	indirect	Dynamin- and Rab5a-mediated Endocytosis of the NK1R Are Required for Resensitization of SP-induced Ca^{2+} Mobilization-- To examine resensitization, cells were incubated with 10 n M SP or vehicle for 10 min, washed, and challenged with 10 n M SP at 0-180 min after washing.	11306580:10246
RARRE S2	retinoic acid receptor responder (tazarotene induced) 2		?	indirect	Chemerin was shown to promote calcium mobilization and chemotaxis of immature DCs and macrophages in a ChemR23-dependent manner.	14530373:6
RGN	regucalcin (senescence marker protein-		+	direct	Regucalcin may have a role as regulatory protein for calcium homeostasis in liver cells., Furthermore, the 45 Ca^{2+} uptake by	1618342:8, 9089642:9, 2632051:3,

	30)		+		the basolateral membranes was clearly increased by the presence of regucalcin (10(-7) and 10(-6) M)., Regucalcin (2.0 microM)-induced retardation of 45 Ca^{2+} uptake was prevented by the presence of calmodulin (5 micrograms/ml)., Regucalcin is involved in maintenance of calcium homeostasis due to the activation of Ca^{2+} pumping enzymes in the plasma membrane., The regucalcin-enhanced ATP-dependent 45 Ca^{2+} uptake by the plasma membrane vesicles was completely inhibited by the presence of NEM (5.0 mM) or digitonin (0.04%). <more data available.>	18157649:0, 9062904:5, 9450663:1, 2049806:1, 14568916:10:
RGS7	regulator of G-protein signaling 7		+	indirect	Here, we demonstrate that RGS7 is a potent GAP *in vitro* on G(alpha i1), and G(alpha o) heterotrimeric proteins and that RGS7 acts to down-regulate G(alpha q)-mediated calcium mobilization in a whole-cell assay system using a transient expression protocol., It was reported that RGS7 inhibited Galpha q -coupled calcium mobilization (30)., Both the RGS domain of RGS7 (17, 18) and full-length RGS7 and RGS7.Gbeta5 (11) inhibit Galphaq-mediated calcium mobilization in transfected cells, suggesting that RGS7 also acts on Galphaq. <more data available.>	9572280:1, 10092682:10 12670932:10: 10840031:10: 15496508:10: 12140291:10:
RIT2	Ras-like without CAAX 2		+	indirect	These results suggest that Rit and Rin define a novel subfamily of Ras-related proteins, perhaps using a new mechanism of membrane association, and that Rin may be involved in calcium-mediated signaling within neurons., These data suggested that dominant negative Rin or siRNA of Rin inhibit calcium-mediated neurite outgrowth and that Rin may be involved in the calcium-mediated signaling pathways leading to neurite outgrowth.	8824319:12, 110003064:1: 4
RRAS	related RAS viral (r-ras) oncogene homolog	+	+	indirect	R-Ras alters Ca^{2+} homeostasis by increasing the Ca^{2+} leak across the endoplasmic reticular membrane.	16469868:10

SELL	selectin L			-		Although L-selectin-induced Ca^{2+} mobilization has been reported to be tyrosine phosphorylation-dependent (28), we show for the first time that tyrosine phosphorylation is required for E-selectin induced signaling.	10748085:10184
SYT1	synaptotagmin I			+	indirect/downstream	Based on previous genetic and structural studies, synaptotagmin-1 was proposed to trigger fast synchronous neurotransmitter release upon Ca^{2+} influx by binding to SNARE complexes, displacing complexin, and coupling the SNARE complex to phospholipids (38).	18308938:10144
TBXA2R	thromboxane A2 receptor			?	indirect	The results clearly indicate that in human intrapulmonary artery, there are TP receptors coupled with phospholipase C activation and that TP receptor-mediated Ca(+2)-mobilization is in part nifedipine- and nitroglycerin-resistant, but forskolin-sensitive., The effect of BM-613 on intracellular signaling by the TP isoforms was investigated by comparing its effect to BM-573 in intracellular calcium mobilization mediated by TPa and TPß stably overexpressed in HEK 293 cells, in response to the selective TXA 2 mimetic U-46619. <more data available.>	8950308:5, 15626721:10185 16156795:10007 107101045:10008

TNNT2	troponin T type 2 (cardiac)		-		The effect of cTnT and/or TnI isoforms on the rate of Ca^{2+} removal from either the high or low affinity Ca^{2+}-binding sites on cTnC has not been examined previously. cTnT had a greater effect on the kinetics of the Ca^{2+} dissociation rate from site II of cTnC in the cTnI.cTnC complex than the ssTnI.cTnC complex (Figs. 7 and 8). cTnT has not been shown previously to alter the kinetics of Ca^{2+} release from site II of cTnC.	15358779:102
TP53	tumor protein p53		+	indirect	B, p53-induced ER calcium release is reduced by BIK knockdown.	15809295:101
VSNL1	visinin-like 1		+	indirect/downs	VILIP-1 modulates the surface expression	164103349:10

					tream	and agonist sensitivity of the in response to changes in the intracellular Ca $^{2+}$ concentration (Lin *et al.,* 2002).	5
WAS	Wiskott-Aldrich syndrome (eczema-thrombocytopenia)			-		A major role of WASp in lymphocyte functions has been implied by the various defects observed in the WAS Th cells, *e.g.*, decreased proliferation, decreased calcium influx, and impaired IL-2 secretion (8, 12, 13).	15728466:10024
ZAP70	zeta-chain (TCR) associated protein kinase 70kDa			-		Calcium mobilization is increased in CD3-cross-linked Syk hi /ZAP-70 - T cells., C, Syk and Zap-70 are involved in Ly-49D-mediated calcium mobilization in Jurkat T cells., Studies in a variety of cellular systems have demonstrated that ZAP-70 is required for both receptor-mediated calcium mobilization and Ras activation (38, 39, 50, 54)., Lck is crucial for the initiation of the tyrosine kinase cascade (36, 37), while ZAP70 is required for PLC?1 phosphorylation and Ca $^{2+}$ mobilization in T lymphocytes (35). <more data available.>	10748099:10133 10553049:10135 9566904:10254, 10452976:10204 10979964:10216 15514014:10123

CHAPTER 3

Development of Mechanistic Model for Drug-Induced Cholestasis and its Applications for Drug Development

Nikolai Daraselia, Pat Morgan and Anton Yuryev[*]

Ariadne Genomics Inc., Rockville, MD, USA

Abstract: We describe construction of mechanistic model for drug-induced cholestasis using information available from ResNet and ChemEffect knowledge networks in Pathway Studio software. We first developed the mechanistic model using information about protein targets of the cholestasis-inducing drugs. We then expanded the model by incorporating knowledge about protein functional annotation, protein homology, canonical pathways and pathway reconstruction. The expanded model provides a toxicity mechanism for 81% of the drugs known to induce cholestasis *vs.* 58% of the drugs explained by the original model. Using the model we suggest that FGF19 secreted proteins are the biomarker for drug-induced cholestasis. In this discussion we suggest how the mechanistic model can be used for predicting cholestasic risk for new compounds, how it can be used for development of biomarker panel to monitor cholestasis risk in patient during drug therapy, and how the model can be used in personalized medicine for evaluating patient predisposition for cholestasis.

Keywords: Bioinformatics, cholestasis, CYP3A4, cytochrome p450, liver toxicity, toxicology, pathway analysis, bile acids, ABC transporters, biomarkers, mechanistic model, MDR1, drug metabolism, pathway studio.

INTRODUCTION

Drug toxicity is the leading cause for dismissing lead compounds during drug development. It also is a frequent reason for the withdrawal of drugs from clinical trials and subsequently from the market, which usually is followed by huge legal liability fines and compensation payments to injured patients. Liver toxicity is the major type of drug-induced toxicity due to the primary role of the liver in xenobiotic chemical metabolization and clearance from the patient's body. Drug metabolization in liver cells is typically described by three phases: oxidation by cytochrome p450 enzymes in the hepatocytes' microsome, conjugation with

*Address correspondence to Anton Yuryev: Ariadne Genomics Inc., Rockville, USA; Email: anton@ariadnegenomics.com

Anton Yuryev and Nikolai Daraselia (Eds)

various hydrophilic anionic groups causing drug inactivation, and the efflux of conjugated and inactivated drugs by the family of multi-drug resistance ABC transporters into bile destined for the intestine for excretion from the body [1].

When a drug or its metabolite inhibits one or several enzymes involved in its metabolism in the liver, it causes accumulation of the drug in the liver and subsequently can cause one or several types of liver toxicity since the same enzymes are also involved in normal liver physiology to digest and detoxify food. Cholestasis is a common type of liver toxicity resulting in the inability to secrete bile [2]. In this article we use cholestasis as example to demonstrate how knowledge networks comprising of drug-target relationships can be used in combination with biological association network to build a mechanistic model for drug toxicity. We used the ChemEffect knowledge base from Ariadne Genomics [3] as a source of drug-target relationships and the ResNet 7 database as a knowledge base for molecular interactions.

Cholestasis can have several causes: ingesting drugs and other chemicals that induce cholestasis, infection causing inflammation and subsequent liver damage, physical damage to the bile duct, and rare genetic mutations. Patients with hereditary cholestasis die at a very young age, making hereditary cholestasis a rare disorder [4]; however, a large number of patients can be genetically pre-disposed to cholestasis due to common SNPs. Unrelated cholestasis causes may synergize or antagonize in a single patient: a patient genetically pre-disposed to cholestasis can develop cholestasis when he or she uses cholestasis-inducing drugs to treat other common diseases such as cancer. On the other hand, patients who are genetically resistant to cholestasis or resistant to infection-induced inflammation can better withstand therapy that includes cholestasis-inducing drugs. This article focuses on building a mechanistic model for drug-induced cholestasis. The same approach may be reused for building a model for inflammatory cholestasis, which is another major cause for this toxicity [5].

Building the model for drug-induced cholestasis is detailed in the *Results* section. It involved the following major steps:

1. Finding proteins targets responsible for development of cholestasis after their inhibition by a drug.

2. Connecting found proteins with regulatory relationship to build the draft model containing only targets known to cause cholestasis.

3. Expanding the draft model to include more proteins in order to enable the model to have better explanatory and predictive power and to enhance model detalization.

4. Validating the model for drug-induced cholestasis with literature data about hereditary cholestasis.

The *Discussion* section of this article shows how the model can be used for drug and biomarker development in personalized medicine.

ABBREVIATIONS

PXR - Pregnane X receptor

FXR - farnesoid X receptor

CAR - constitutive androstane receptor

PPARA - peroxisome proliferator activated receptor alpha

RXR - retinoid-X receptor

HNF1A - hepatic nuclear factor-1-alpha

HNF4A - hepatic nuclear factor-4-alpha;SHP - small heterodimer partner

FGF19 - fibroblast growth factor 19

OSTalpha - organic solute transporter alpha

OSTbeta - organic solute transporter beta;

RESULTS

Building the Model for Drug-Induced Cholestasis Using Knowledge Networks: To find proteins involved in development of cholestasis we first found 62 drugs in the ChemEffect 2.0 database known to induce cholestasis as described in the *Methods* section. We then found 401 protein entities inhibited by at least two drugs inducing cholestasis hypothesizing that cholestasis is mainly caused by inhibition of off-target proteins involved in bile acid metabolism. We also hypothesized that drugs induce cholestasis through one or more common mechanisms and therefore only proteins inhibited by at least two drugs were selected for model construction. Selecting only common targets also allowed elimination of drug therapeutic targets and focused only on drug off-target effects.

Manual inspection of several protein targets revealed that many cholestasis-inducing drugs were anti-cancer drugs. Hence they shared not only off-target proteins that could be linked to the common toxicity but they also could have common therapeutic targets. To further improve the relevance of protein components to the cholestasis model we selected only 56 proteins out of 401 common targets due to their known role in cholestasis. The search for these proteins in the knowledge networks is detailed in the *Methods* section.

Manual inspection of the functional annotation from 56 selected proteins revealed that they can be classified into two functional categories: proteins directly involved in bile acid metabolism, and their regulators such as transcription factors and hormones. We focused on building a draft of a cholestasis model from only 21 proteins known to have a direct role in bile acid metabolism. These proteins can be further classified into bile acid importers into hepatocytes, bile acid synthesis and conjugation enzymes, and bile acid exporters from hepatocytes (Fig. 1). This is consistent with the current view of bile acid circulation which postulates that hepatocytes preferentially import damaged unconjugated bile acids and export conjugated bile acid [6], while compensating for the lack of circulating bile by its synthesis from cholesterol.

Next, we found the principal transcription factors regulating levels of 21 proteins involved in bile acid circulation and metabolism using sub-network enrichment analysis [7]. Seven transcription factors found as best sub-network enrichment hits included six nuclear receptors and hepatic nuclear factor-1 in the following order of statistical significance: PXR, FXR, CAR, PPARA, RXR, HNF1A, and SHP. HNF1A functions in liver growth and differentiation and therefore was excluded from the list of bile acid homeostasis regulators. Among the remaining nuclear receptors we found that only FXR and PXR could directly bind to bile acids, which leads to the activation of their transcription activities [8,9]. Additionally, only FXR and PXR can activate the expression of proteins involved in bile acid conjugation and secretion and at the same time repress expression of genes involved in bile acid synthesis and import into hepatocytes (Fig. **2**). We concluded that, based on their regulation pattern, FXR and PXR appear to act as principal internal sensors of bile acid concentration in hepatocytes. If there is too much bile in the hepatocytes, FXR and PXR induce conjugation of bile acids to increase their secretion from the cell, suppressing bile acid synthesis and import at the same time. If there is a lack of bile inside hepatocytes, FXR and PXR derepress synthesis and import of bile acids while slowing their conjugation and secretion.

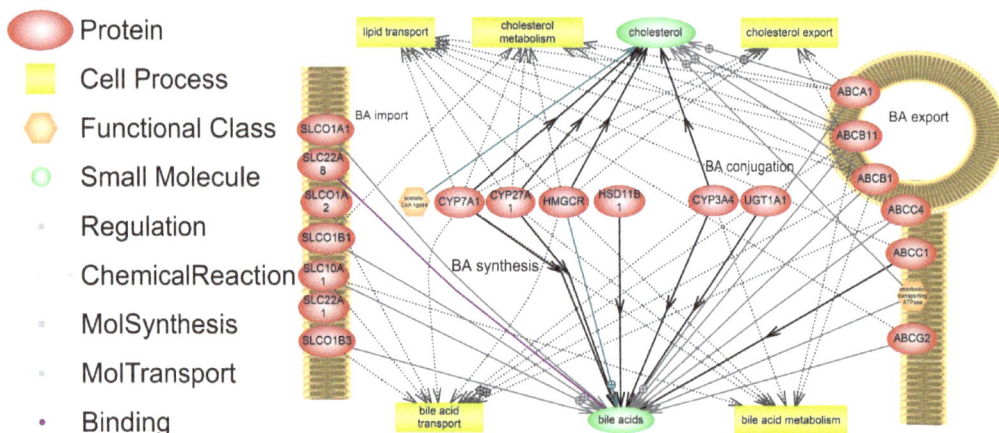

Figure 1: A screenshot of Pathway Studio showing 21 proteins chosen for a draft of the drug-induced cholestasis model.

In order to regulate bile production on the level of the entire organ, the hepatocyte cell must signal neighboring liver cells about the current internal bile acid concentration state. Since neither FXR nor PXR can be exchanged between hepatocytes, autocrine signaling through secreted ligands must be used to propagate a signal of internal bile acid concentration through the liver. To find such a secreted signaling molecule we inspected common expression targets downstream of FXR and PXR in the ResNet 7 database and identified only one secreted protein among them—FGF19. Publications supporting PXR->FGF19 Expression regulation suggest that FGF19 is the principal mediator of suppression of bile acid synthesis by PXR and FXR [8]. Reading other statements from the literature supporting FXR->FGF19 in the ResNet 7 database revealed that the SHP transcription factor is the principal mediator of FGF19 autocrine signaling that suppresses expression of CYP7A1 and CYP27A1 - two enzymes catalyzing rate-limiting steps in bile acid synthesis [9, 10].

The final draft of the cholestasis model containing known targets of cholestasis-inducing drugs and their transcriptional and autocrine regulators is shown in Fig. **2**. It contains targets for 37 out of 62 drugs known to induce cholestasis. Thus, it provides a mechanism of action for 58% of the drugs. Among 26 cholestasis-inducing drugs not explained by the model, 17 drugs did not have any targets known to play a role in cholestasis and 9 drugs inhibited targets known to play a role in cholestasis but do not directly function in bile acid circulation or metabolism.

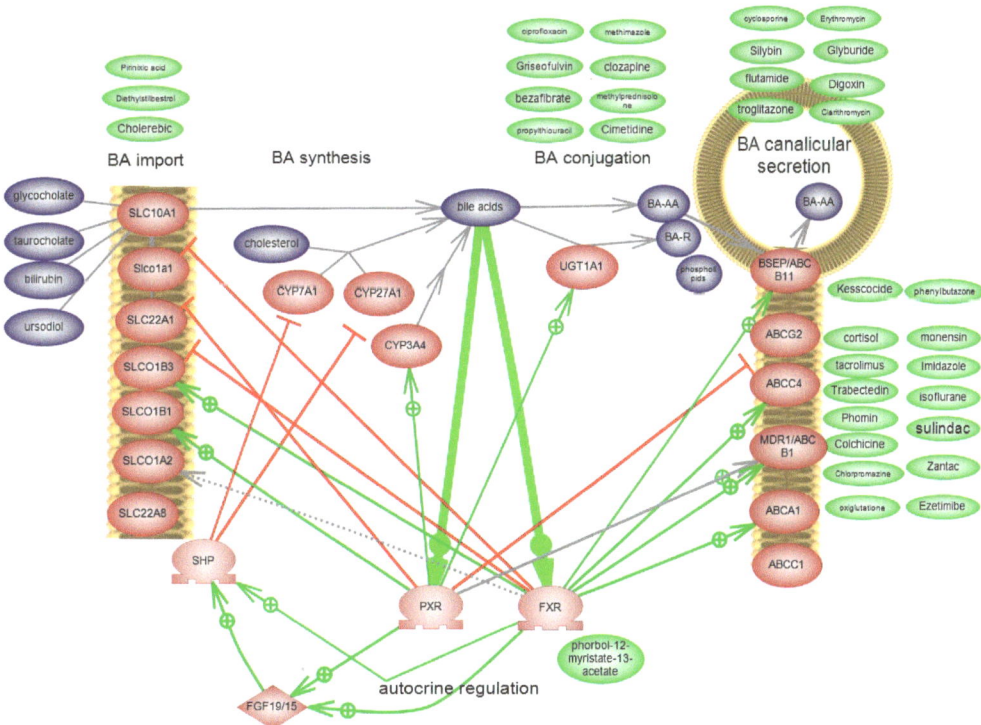

Figure 2: A draft cholestasis model consisting of the proteins directly involved in bile acid circulation and metabolism that are inhibited by cholestasis-inducing drugs. The model shows bile acid (BA) homeostasis circulation through the hepatocyte. These liver cells preferentially import damaged and unconjugated bile acids and secrete into the bile duct repaired bile acids conjugated (BA-AA) with aminoacids: taurine and glycine. Bile acids also can be modified by other anionic

chemical groups (BA-R) such as glucoronate or sulfate that also promotes their excretion from hepatocytes. The 37 drugs explained by the model are shown next to their respective targets.

The draft cholestasis model shown in Fig. **2** contains indirect FGF19 autocrine regulation links, an incomplete metabolic pathway for bile acid synthesis, and no details about which bile acids can activate PXR and FXR regulators. To enhance the predictive power of our cholestasis model we decided to expand it by adding detailed molecular mechanisms behind indirect links and by adding other proteins involved in bile acid circulation and metabolism. Even though such an expanded model contains proteins that are not known targets of cholestasis-inducing drugs, the inhibition of these proteins by a new drug may potentially cause cholestasis through inhibition of bile acid circulation.

Expanding the Cholestasis Model Using Protein Functional Homology: Functional homologs of the proteins involved in bile acid circulation were identified using the BLAST program in order to find proteins with amino acid sequence similarity higher than 50% to transporters in the draft cholestasis model shown on Fig. **2**. Using this approach we added the following proteins: SLC10A2 as the homolog of SLC10A1 transporter; SLCO1A2, SLCO4C1 as homologs of SLCO1B1; CYP7B1, CYP8B1 as homologs of CYP7A1; ABCB5 as homologs of ABCB11/BSEP; ABCC3, ABCC6, and ABCC2 as homologs of ABCC1. A subsequent inspection of protein annotation confirmed the role of these proteins in bile acid transport or synthesis.

To add more functional homologs we have selected transporters reported to regulate bile acid transport as described in the *Methods* section. This approach added OSTalpha and OSTbeta transporters to the model. These two proteins have a sequence similarity below 50% as compared to other bile transporters and therefore were missed by the sequence similarity search.

Expanding the Model by Adding Bile Acid Metabolism Pathway: A bile acid metabolism pathway is readily available from the Ariadne metabolic pathway collection in ResNet 7 database, and therefore it was copied "as is" to extend the model. CYP7A1, CYP27A1 and CYP3A4 are the only enzymes of bile acid synthesis inhibited by cholestasis-inducing drugs. It is known that both CYP7A1 and CYP27A1 catalyze rate limiting steps in the bile acid synthesis [9, 10]. To find other rate-limiting steps in bile acid synthesis we looked for the most regulated enzymes in the bile acid synthesis pathway. This was done by building a transcription regulatory network as described in the *Methods* section. Results of the analysis are shown in Table **1**.

Table 1: Protein connectivity in the expression regulatory network for bile acid synthesis. In-Degree is the number of relations where protein is a target. Out-degree is the number of relations where a protein is a regulator.

Protein	In-Degree	Out-Degree	Function
CYP7A1	13	0	Bile acid synthesis enzyme
CYP27A1	10	0	Bile acid synthesis enzyme
CYP8B1	9	0	Bile acid synthesis enzyme
CYP7B1	5	0	Bile acid synthesis enzyme
BAAT	4	0	Bile acid conjugation enzyme
SLC27A5	3	0	Bile acid synthesis enzyme
SP1	0	9	Transcription factor
HNF4A	0	8	Transcription factor
SREBF1	0	6	Transcription factor
SREBF1	0	6	Transcription factor
PPARA	0	4	Transcription factor
FXR	0	4	Transcription factor
PXR	0	3	Transcription factor
RARA	0	3	Transcription factor
SHP	0	3	Transcription factor

CYP7A1, CYP27A1 were found to be the most regulated enzymes. CYP8B1 and CYP7B1- proteins added to the expanded model as sequence homologs of CYP7A1 and CYP27A1—were also highly regulated. Finding that these proteins are hubs in the expression network further validated adding these proteins to the cholestasis model.

Expanding the Model by Reconstructing FGF19 Autocrine Signaling Pathway in Hepatocytes: Since FGF19 is a secreted ligand, it requires a signaling pathway consisting of receptor and signal transduction molecules conducting the signal to metabolic enzymes for repression of bile acid synthesis. Because the FGF19 pathway was not described in the literature and was not available in the Ariadne canonical pathway connection, we reconstructed it using data from the ResNet 7 database. Pathway reconstruction is described in detail in the *Methods* section. In brief, we have used several approaches including finding statements about pathway intermediates in scientific literature, regulome approach [14], and conservative signaling blocks identified by pathway alignment with canonical FGFR1/2 pathways. The resulting pathway is fully consistent with experimental observations from multiple publications (Fig. **3**). The FGFR4 protein was supported by 14 references in the ResNet 7 database as the FGF19 receptor.

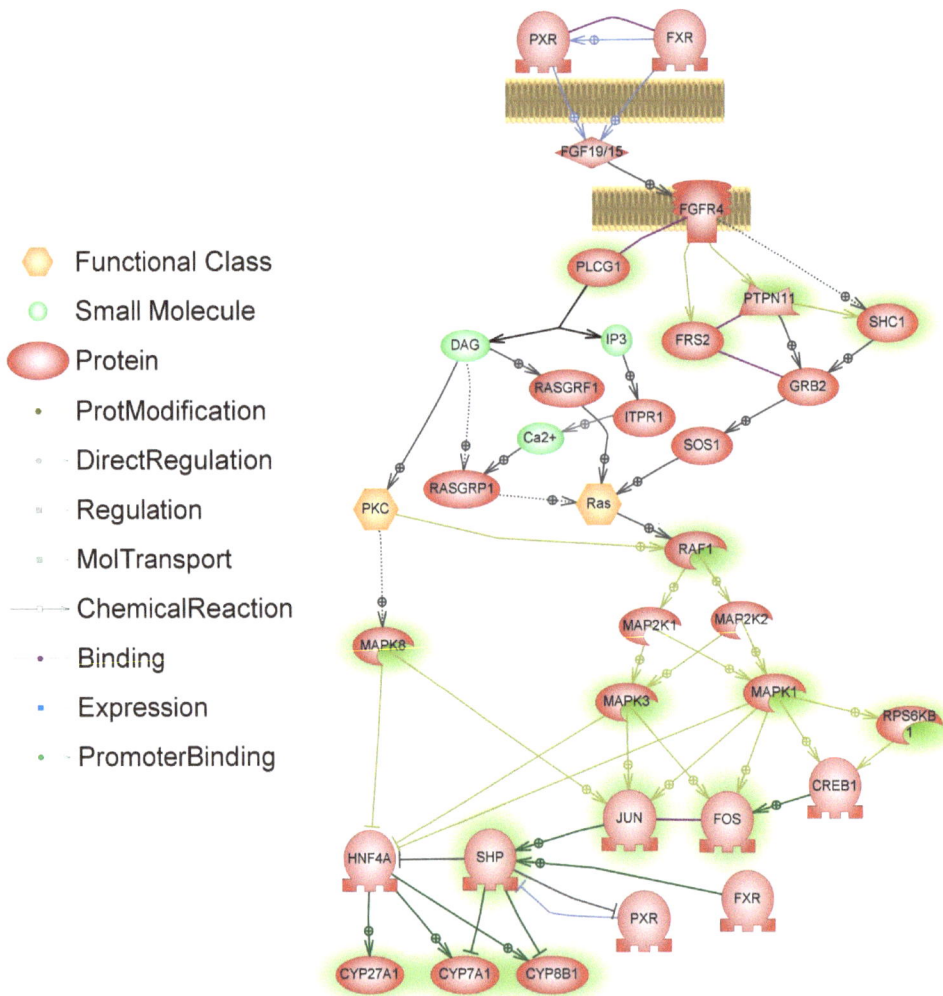

Figure 3: The predicted FGF19 signaling pathway repressing expression of enzyme for bile acid synthesis. Proteins from the FGF19-FGFR4 regulome are highlighted in green. These proteins are

reported as being activated by either FGF19 or FGFR4. Other proteins were added to the pathway because of their similarity to the canonical FGFR1 signaling pathway available from the Ariadne signaling pathway collection in the ResNet 7 database. They represent conserved signaling blocks that exist in multiple signal transduction pathways. HNF4A was added to the pathway because it is the major transcriptional regulator of bile acid synthesis enzymes (Table **1**).

The expanded cholestasis model is shown in Supplementary Fig. **1**. This model contains 51 proteins, including members of 21 functional classes representing enzymatic steps in bile acid synthesis and 57 chemicals: unconjugated and conjugated bile acids, and intermediate steps of bile acid synthesis. The expanded model contains 20 proteins from the draft cholestasis, which are known targets of cholestasis-inducing drugs. Various applications for the model are described later in this chapter. In order to prove that the expanded model has higher predictive power we measured how many cholestasis-inducing drugs regulate at least one protein in the model. We found that the expanded model is regulated by 51 drugs, *i.e.*, by 81% of all drugs known to induce cholestasis in the ResNet 7 database. This is a 23% increase in power, as compared to the draft cholestasis model made only from protein targets linked to cholestasis in the ResNet database. The increase occured because of the addition of proteins involved in bile acid circulation and metabolism but were not linked to cholestasis in the literature. Most drugs were found to regulate CYP3A4, ABCB1/MDR1, SCLCOB1/OATP2 and ABCB11/BSEP transporters (Supplementary Table **1**). These results are discussed later in this chaper.

Selecting FGF19 as a Highly-Specific Functional Toxicity Biomarker Using the Cholestasis Model: FGF19 or its mouse homolog known as FGF15 was previously established as enterohepatic autocrine signal regulating bile acid homeostasis [18]. According to our model, the majority of drugs inducing cholestasis (35 out of 62) inhibit bile acid efflux transporters. Twenty-four of these drugs are being metabolized by CYP3A4, causing its inhibition. Inhibiting bile acid secretion by a drug increases bile concentration inside hepatocytes, activating PXR-FXR transcription factors. In normal circumstances this would lead to increased expression of bile acid exporters, decreased expression of bile acid importers, and downregulation of bile acid synthesis locally through expression of SHP and globally through FGF19 signaling. All these molecular events allow hepatocytes to clear out the excess of bile and drug by overproducing bile acid exporters inhibited by a drug, thus allowing healthy bile acid circulation, albeit at possibly lower rate.

CYP3A4 and its paralogs CYP3A5/6 are the most abundant cytochrome P450 in the liver comprising 30-40% of the total CYP content in the human adult liver and small intestine [19]. They perform phase I degradation of more than 50% of all known drugs oxidizing the drug. The oxidized drug molecule is then glucaronidized, sulfated, acetylated, glutathionated or undergoes some other type of hydrophilic conjugation in phase II that facilitates its export from the cell through multidrug resistance transporters [20]. Many drugs bind CYP3A4 but some are difficult to metabolize, causing inhibition of CYP3A4 enzymatic activity. Due to the major role that CYP3A4 plays in drug metabolism there is a significant number of drugs known to inhibit CYP3A4. For example, the ChemEffect 2.0 database has 339 drugs reported as CYP3A4 inhibitors. The liver responds to the inhibition of CYP3A4 by over-expressing CYP3A4. This is achieved through activation of either PXR or a glucocorticoid receptor [21] by the accumulating drug that can act as ligand for these nuclear receptors [22, 23]. PXR can be further activated by increased production of cholestane-3,7,12,25-tetrol from bile acid intermediate trihydroxycoprostane by CYP3A4 [24]. Cholestane-3,7,12,25-tetrol can also activate FXR [12] and can be accumulated in hepatocytes due to inhibition of bile acid secretion by a drug. In summary, drugs that are capable of inhibiting CYP3A4 and, at the same time, inhibiting bile acid exporters tremendously activate PXR-FXR transcription factors through two mechanisms: by the increase in the concentration of bile and its intermediates in the hepatocytes and by direct binding to PXR or the glucocorticoid receptor. The inability of CYP3A4 to metabolize these drugs effectively causes an increase in drug concentration inside hepatocytes which further blocks bile acid excretion and activates PXR and FXR even more. Our model for drug-induced cholestasis suggests that the transcriptional hyperactivity of PXR-FXR proteins is the main cause of shutting down bile acid synthesis in the cholestatic liver, which is mediated by FGF19 autocrine signaling. Therefore, FGF19 levels must be indicative of the cholestasis status of a patient's liver. FGF19 is a secreted ligand that can be easily assayed in the blood of a patient. Based on its function and properties we propose that FGF19 can be used as a functional clinical biomarker to measure the cholestatic status of patients treated with cholestatic drugs. FGF19 over-expression in cholestasis was recently confirmed experimentally [25].

FGF19 also appears to be the most specific biomarker for cholestasis. According to the Ariadne Disease*FX* database 1.0 it is over-expressed in only three disease conditions: in the end stage of renal disease [26], in lung squamous cells and in colon adenocarcinomas [27]. For comparison, the most specific biomarker described for cholestasis in the literature—serum proenkephalin—was shown as a biomarker in 20 other diseases (Supplementary Table **2**). Normal FGF19 levels can be measured prior to drug therapy. FGF19 levels can be then monitored during therapy to anticipate the onset of cholestasis.

Validating the Cholestasis Model with Genetic Data on Hereditary Cholestasis: To further validate our drug-induced cholestasis model we compared it with proteins linked to hereditary cholestasis. We wanted to know if mutations in genes involved in drug-induced cholestasis are linked to hereditary cholestasis. Genes linked to hereditary cholestasis were found using the Ariadne Biomarker cartridge as described in the *Methods* section. Out of 22 genes with mutations known to cause cholestasis, 10 genes belonged to our cholestasis model. The CYP7B1 gene is linked to hereditary cholestasis but is not part of our cholestasis model. It belongs to the alternative bile acid synthesis pathway. This pathway is also controlled by PXR-FXR-SHP transcription factors because it contains CYP27A1 and CYP7A1. Therefore, drugs inducing cholestasis must repress alternative bile synthesis as well.

METHODS

Building the Cholestasis Model: To find drugs that induce cholestasis we first found the disease entity *cholestasis* in the ResNet 7 database updated with ChemEffect 2.0 data and then expanded the *cholestasis* entity upstream filtering for Small Molecules and for Regulation with Effect positive. We then manually removed non-drug small molecules such as bile acids and opioid metabolites known to induce cholestasis when administered at high concentrations.

Protein targets inhibited by cholestasis-inducing drugs were found using the "Add common targets" option in the advanced Build pathway dialog in Pathway Studio. To select only protein targets we filtered for *Protein*, *Functional Class* and *Complex*, and to select for protein inhibited by the drugs we filtered for "Regulation" and "Direct Regulation" relations with Effect negative.

To find protein targets playing a role in cholestasis we first expanded the *cholestasis* entity upstream filtering for *Protein*, *Functional Class* and *Complex* entity and for Regulation with Effect positive or unknown. To find more proteins involved in cholestasis we also expanded the *Cholestasis, Intrahepatic*, and *Cholestasis, Extrahepatic* entities using the same filtering option. These searches yielded a collection of 110 protein entities that was then intersected with 401 proteins inhibited by at least two drug inducing *cholestasis* using the "Select Entities on clipboard" option in Pathway Studio. After manual curation by inspecting their functional annotation and by removing highly generic and therefore uninformative functional classes such as *Cytokines* and *Monooxynases,* we selected 58 protein entities that were known to be inhibited by at least two drugs inducing cholestasis.

To find transcription factors regulating the expression of proteins involved in bile acid metabolism we used the command "Find sub-networks enriched with selected entities" with the option *Expression targets*. Reading supporting sentences for FXR and PXR expression sub-networks revealed that these proteins can bind bile acids and form heterodimer on the downstream gene promoters. To find secreted proteins regulated by FXR and PXR we used the option "Find common targets" filtered by *Expression* relations. Among the 17 targets shared by FXR and PXR, only the FGF19 protein was also a secreted protein. This was evident from the FGF19 node rhomb shape indicating that it was classified as ligand and later was confirmed by inspecting proteins annotation of all 17 common targets. Other FXR and PXR targets were either transporters or enzymes involved in bile acid synthesis.

Expanding the Cholestasis Model with Bile Acid Biosynthesis Metabolic Pathway: A BLASTp of the entire human proteome against itself was performed using a program from NCBI and a full-length aminoacid sequence similarity was calculated using BLAST output as described previously [13]. All paralog pairs were imported into the Pathway Studio database as a new relation type called "Paralog" annotated with sequence similarity. As a result, finding paralogs of proteins involved in bile acid circulation and synthesis amounted to finding neighbors in the Paralog network with a 50% similarity cutoff.

To find other bile acid transporters that could have been missed by sequence homology search we found all proteins regulating the *Bile acid transport* Cell Process in Pathway Studio and then selected only the proteins annotated as transporters.

To find rate-limiting steps in bile acid synthesis we have collected all *Protein* members of *Functional Classes* involved in bile acid synthesis and searched for common regulators in the *Expression* network. This approach yielded 12 transcription factors known to regulate 16 proteins involved in bile acid synthesis. The most regulated proteins were found by sorting all proteins in expression regulatory network by In-Degree. The results of this analysis are shown in Table **1**.

Expanding the Cholestasis Model by Reconstruction of the FGF19 Signaling Pathway: Inspecting references supporting the FGF19->CYP7A1 Expression relation identified FGFR4 receptor, small heterodimer partner (SHP) and JNK pathway as an intermediate steps for FGF19 signaling [15, 16, 17]. To find other members of FGF19 signaling pathway we built FGF19-FGFR4 regulome by finding all proteins downstream of either FGF19 or FGFR4 in *Regulation*, *DirectRegulation*, *ProtModification* or *Binding* networks. Signal transduction proteins from FGF19-FGFR4 regulome were then connected with physical interactions: *Binding*, *DirectRegulation*, *ProtModification*. The remaining gap between JNK signaling and FGFR4 adaptor proteins was closed by copying conserved signaling blocks from the FGR1->RUNX2 canonical pathway from the Ariadne pathway collection: GRB2-SOS1-RAS block was added to activate RAF kinase, and -gamma-PKC block was added to activate JNK1 (MAPK8).

Measuring Biomarker Specificity: The Ariadne Disease*FX* 1.0 contains more than 135,000 statements about changes in protein states in various disease conditions. Statements were searched using the MedScan technology in more than 18 mln PubMed abstracts and more than 500,000 full-text articles. Biomarker specificity is defined as the number of diseases where a protein changes its abundance. We used following relations extracted by the MedScan Biomarker cartridge to find abundance biomarkers: *ExpressionChange*, *AbundanceChange*, and *ActivityChange*.

To compare FGF19 biomarker specificity with the specificity of other cholestasis biomarkers we first found all biomarkers reported in the literature that increase their abundance during cholestasis. This was done by expanding the entity *cholestasis* in the Pathway Studio database downstream to find *Protein*, *Functional Class* and *Complex* entities linked to cholestasis by *ExpressionChange*, *AbundanceChange*, and *ActivityChange* relations with Effect positive or unknown. We then expanded all found biomarkers upstream to find Disease entities linked to cholestasis biomarkers by *ExpressionChange*, *AbundanceChange*, and *ActivityChange* relations with Effect positive or unknown. The biomarker specificity in this network is In-Degree of all cholestasis biomarkers. In-Degree is readily calculated in Pathway Studio for all entities in the network and available in the Entity table view of the network. The specificity of all known cholestasis biomarkers is shown in Supplementary Table **2**. FGF19 biomarker specificity was measured using the same method.

Finding Proteins Genetically Linked to Hereditary Cholestasis: We used the Ariadne Disease*FX* database to expand *cholestasis* disease entity downstream in the *GeneticChange* knowledge network extracted by the Ariadne Biomarker cartridge. This approach yielded 31 proteins, which genetic changes were linked to cholestasis by statements in the scientific literature. A closer inspection of the literature statements supporting *GeneticChange* relations revealed that genetic changes of TNF, IL1B, and MET receptor were linked only to the severity of cholestasis rather than its cause. Statements regarding seven more proteins related to protein state alternation rather than genetic alternation. These proteins were excluded from analysis, leaving 21 proteins for comparison with the drug-induced cholestasis model (Supplementary Table **3**).

DISCUSSION

Model for Drug-Induced Cholestasis: Our drug-induced cholestasis model was built using data from the ChemEffect 2.0 database. To further support our findings that CYP3A4 metabolized drugs inhibiting bile export cause cholestasis, we applied data from the most recent ChemEffect 3.0 database to our model. We found occurences of 175 drugs inducing cholestasis in ChemEffect 3.0. One-hundred thirty-eight of them (79%) were found to regulate at least one protein in our cholestasis model built using ChemEffect 2.0 data. Furthermore, we found that 95 drugs regulate bile acid exporters and 60 of them also regulate CYP3A4.

Thus 34% of the drugs known to induce cholestasis in ChemEffect 3.0 are also known to regulate at least one bile acid exporter and CYP3A4. The cholestasis model proposed in this paper is similar to the recently proposed model for drug cholestasis [28] which was built independently without the Pathway Studio knowledge base.

Our model explains why cholestasis is one of the most common drug-induced toxicities. The liver uses the same molecular machinery to maintain bile acid homeostasis, to detoxify food, and to remove drugs. Most drugs or their metabolites are cleared from the liver *via* the family of multidrug resistance ABC-transporters. MDR1/ABCB1 is the major drug exporter linked to more than 1,000 drugs in the ChemEffect 3.0 database. It is also the second major exporter of conjugated bile acids after ABCB11/BSEP. MDR1 is capable of substituting BSEP in BSEP knockout transgenic mice [29]. BSEP and MDR1 are structural homologs: therefore, drugs capable of inhibiting MDR1 also may inhibit BSEP but with different affinities.

The physiological role of CYP3A4 and other cytochrome P450s is clearing liver from toxic bile acids and other chemicals produced by bacterial flora or ingested with food [30]. Most drugs are designed to withstand a first-pass effect caused by drug degradation by cytochrome P450s in the liver. If a drug cannot resist liver metabolization, its effective bloodstream concentration drops soon after digestion, preventing the drug from reaching a target tissue at the effective dose. This phenomenon is called the first-pass effect. Since CYP3A4 is the major enzyme for drug degradation, the task of avoiding the first-pass effect directs drug development towards selecting compounds that can reversibly bind to CYP3A4 but not being metabolized, *i.e.*, competitive CYP3A4 inhibitors. To avoid the accumulation of a non-degradable drug in the liver, the drug absolutely must be exported easily by multidrug resistance transporters back into the gut to allow further absorption into the bloodstream and to travel further to reach the target tissue.

Bile acid exporters and multidrug resistance proteins belong to one family of ABC cassette glycoproteins that use ATP energy to transport various chemicals from hepatocytes and other tissues. They have two translocation and two nucleotide-binding domains. Most drugs bind to the translocation domain but some can also bind to the nucleotide-binding domain [38, 39]. The increased interest of the pharmaceutical industry in kinase inhibitors designed to bind to the ATP-binding domain of proteins kinases should increase the number of drugs capable of binding to the nucleotide-binding domain as well. MDR1 and its paralogs are often over-expressed in cancer cells [40], causing drug-resistance in cancer. The pursuit for better anti-cancer drugs selects chemicals inhibiting MDR proteins in order to achieve a higher anti-cancer efficacy. Thus, former and current drug development efforts are biased towards creating MDR and CYP3A inhibitors and increase the probability of developing drugs with cholestasis as a side effect: there are 387 drugs known to inhibit MDR1 and 404 drugs known to inhibit CYP3A4 in the ChemEffect 3.0 database, proving this point. High efficacy drugs are usually selected in *in-vitro* experiments, thus ignoring possible side-effects such as liver toxicity. Therefore, side-effects such as cholestasis are only noticed later in drug development either during drug testing in an animal model or in patients. Our model suggests that every candidate drug structure should be tested by virtual docking experiments for possibility to bind and inhibit MDR1 and its paralogs in order to reduce risk of "discovering" post factum its cholestatic potential and in order to alert physicians about its cholestasis risk so that the adverse drug reaction can be averted during drug therapy.

The natural function of FXR is sensing bile acid concentration and regulating the rate of bile acid synthesis in response to changes in bile acid levels [31]. The natural function of PXR in liver is less clear. It is known to be activated by toxic bile acids generated by bacterial intestinal flora such as lithocholic acid [32]. PXR can protect the liver from lithocholic acid-induced hepatotoxicity by inducing CYP3A4 for detoxification [33]. More recently, the vitamin D receptor (VDR) was shown to induce CYP3A4 expression both in the liver and in the intestine in response to lithocholic acid [34]. Regardless of PXR's physiological role, the ChemEffect 3.0 database clearly suggests that PXR is the primary xenobiotic sensor among all nuclear receptors. PXR can be activated by 93 drugs while FXR or VDR are known to be activated only by 5 and 7 drugs respectively according to ChemEffect 3.0. This strongly indicates that induction of PXR should play a central role in drug-induced cholestasis. The PXR protein has a promiscuous ligand-binding domain [35] suggesting that PXR's physiological role is to protect the liver from various toxic products either generated by bacteria or ingested with food.

The activation of PXR alone is insufficient to induce cholestasis. For example, only 11 PXR-activating drugs can induce cholestasis according to ChemEffect 3.0 data. Under normal circumstances, PXR activation by xenobiotics decreases intracellular bile acid synthesis caused by the inhibition of HNF4A by PXR on the CYP7A1 promoter [36], due to over-expression of bile acid exporters, and due to down-regulation of expression bile acid importers. PXR activation also inhibits bile acid synthesis globally in the entire liver because of the inhibition of HNF4A *via* the FGF19-FHFR4 pathway and up-regulations of SHP. All of these processes allow hepatocytes to refocus from bile circulation to drug degradation and clearance. However, if the drug blocks bile acid secretion, the accumulating bile hyperactivates FXR in the cell that already has hyperactivated PXR. Activation of FXR further inhibits bile acid synthesis locally through over-expression of SHP, PXR and globally through further increase in FGF19 expression.

The previous scenario suggests that FGF19 can be used as an indicator for autocrine down-regulation of bile acid synthesis but it does not allow differentiation between healthy liver response due to PXR activation and abberant FXR activation as a result of inhibition of bile secretion by the cholestasis-inducing drug. To detect FXR activation another biomarker that specifically indicates that FXR activation is necessary. We have compared all known targets of FXR and PXR in the ResNet database to find secreted proteins that are regulated only by FXR. We found kininogen 1 (KNG1) as one candidate biomarker for FXR activation [37]. On the other hand, both KNG1 and FGF19 levels can be activated due to FXR activation alone. FXR can be activated by all-trans-retinoic acid and triglycerides [41] due to its interaction with a retionoic acid receptor (RXR) [42] or because the FXR gene is regulated by a peroxisome proliferator-activated receptor gamma [43]. The FXR level also varies due to indivudual genetic variation [44]. Thus the combination of KNG1 and FGF19 may indicate activation of FXR caused by an individual's diet or genotype rather than by drug toxicity without activation of PXR. To further improve an assay for early detection of drug-induced cholestasis, a third biomarker specific to PXR is necessary. Recently gamma-alpha-carboxyethyl hydroxychroman beta-D-glucoside, which is a vitamin E metabolite, was identified as a PXR-specific urine biomarker [45]. In summary, our model suggests using biomarker triplet for measuring cholestatic risk in patients during drug therapy: A blood test to measure the elevation of KNG1 to detect

FXR activation and the elevation of FGF19 to detect autocrine signaling to down-regulate bile acid synthesis, and one urine tcst for gamma-alpha-carboxyethyl hydroxychroman beta-D-glucoside levels to detect PXR activation.

APPLICATIONS OF THE CHOLESTASIS MODEL

1) Estimating Cholestatic Risk for Drug Targets: The ChemEffect data and the model for drug-induced cholestasis allow straightforward calculation of cholestatic risk associated with inhibition of a drug target. The approach is illustrated in Table **2**. We found MDR1 inhibition has a relatively low risk of inducing cholestasis (12%), as compared to BSEP or ABCB4 inhibition (39% and 50% respectively). The risk associated with ABCB4 and ABCC3 inhibition is

Table 2: Cholestasis risk associated with the inhibition of various bile acid transporters in our cholestasis model. The risk was estimated as a proportion of drugs known to induce cholestasis among the total number of known transporter inhibitors. To find drugs inhibiting each transporter we found upstream Small molecule neighbors of the transporter in the ChemEffect 3.0 database in Regulation and DirectRegulation networks filtered for Effect "negative". We then compared all found inhibitors with drugs known to induce cholestasis to calculate cholestatic risk.

Transporter	Specificity	# Inhibitors	# Inhibitors Known to Induce Cholestasis	Cholestatic Risk
ABCB4/MDR3	phospholipids	4	2	0.5
ABCC3	organic anions	5	2	0.4
BSEP	conjugated bile acids	23	9	0.391
ABCC2/Mrp2	organic anions, sulfated and glucoronidate d bile acids	44	10	0.227
ABCC4/Mrp4	organic anions	26	5	0.192
ABCG2	xenobiotics	65	11	0.169
MDR1	xenobiotics	323	38	0.117
ABCC1	organic anions, sulfated and glucoronidate d bile acids	44	4	0.091

most likely over-estimated due to the small number of known inhibitors. However, BSEP inhibition risk is expected to be higher than MDR1 inhibition risk because BSEP is the major bile exporter that has a fivefold higher affinity towards bile acid than MDR1 [29].

2) Predicting Compound Cholestatic Risk: Our cholestasis model suggests that for a compound to cause cholestasis it must inhibit bile acid export and induce PXR activity at the same time. PXR can be activated due to the inhibition of CYP3A enzymes metabolizing more than 50% of all known drugs. Further inspection of PXR targets in the ResNet database revealed that PXR regulates the expression of at least 23 other cytochrome p450s enzymes including CYP2C9, which can metabolize more than 10% of existing drugs. While most new small molecule drugs will activate PXR, only a relatively small subset of them will also block bile acid transporters. Therefore, one way to predict cholestatic risk for a new compound is to perform virtual docking simulations first against BSEP, MDR1 and other bile acid exporters and then against cytochrome p450s regulated by PXR. Pharmacophore models for virtual docking can be calculated based on known structures of mouse MDR1 [46] and homology modeling for human MDR1, BSEP, and other transporters. A pharmacophore model for CYP3A4 and other p450s has already been developed [47,48,49].

Another way of assessing the cholestatic potential of a new compound is by monitoring intracellular bile concentration leading to the activation of FXR. The FXR activation must be concomitant with PXR activation to indicate cholestatic risk. A recently developed *in vitro* assay using sandwich-cultured rat hepatocytes monitoring accumulation of intrahepatocellular bile [50] can be used for measuring activation of FXR and PXR. The activity of these transcription factors can be calculated using sub-network enrichment analysis from the microarray gene expression data [7].

3) Workflow for Using the Cholestasis Model in Personalized Medicine. Our cholestasis model provides the foundation for evaluating cholestatic predisposition in a patient. This may be important in cases when a patient is prescribed with a cholestatic drug. Common SNPs causing a slight reduction in the activity of bile acid exporters or causing hyperactivity of PXR or FXR proteins may allow the development of drug-induced cholestasis at low drug doses tolerated by non-predisposed patients. In cases in which no alternative to cholestatic drug therapy is available, predisposed patients can be prescribed with an anti-cholestatic drug to minimize risk of adverse reaction.

CONFLICT OF INTEREST

Authors do not have any conflicts of interests with respect to chapter content.

ACKNOWLEDGEMENT

We are grateful to Dr. Steven Spanhaak from Johnson & Johnson Inc. for suggestion to build a model for drug-induced cholestatsis and useful discussions.

REFERENCES

[1]   Muntané J. Regulation of Drug Metabolism and Transporters. *Curr Drug Metab.* 2010;10(8):932-45.
[2]   Kleiner DE. The pathology of drug-induced liver injury. *Semin Liver Dis.* 2009; 29(4):364-72.
[3]   Yuryev A, Kotelnikova E, Daraselia N. Ariadne's ChemEffect and Pathway Studio knowledge base. *Expert Opinion on Drug Discovery.* 2009; 4(12):1307-1318.
[4]   Davit-Spraul A, Gonzales E, Baussan C, Jacquemin E. Progressive familial intrahepatic cholestasis. *Orphanet J Rare Dis.* 2009;4:1.
[5]   Kosters A, Karpen SJ. The role of inflammation in cholestasis: clinical and basic aspects. *Semin Liver Dis.* 2010;30(2):186-94.
[6]   Hofmann AF. Biliary secretion and excretion in health and disease: current concepts. *Ann Hepatol.* 2007;6(1):15-27.
[7]   Sivachenko AY, Yuryev A, Daraselia N, Mazo I. Molecular networks in microarray analysis. *J Bioinform Comput Biol.* 2007 Apr;5(2B):429-56.
[8]   Wistuba W, Gnewuch C, Liebisch G, Schmitz G, Langmann T. Lithocholic acid induction of the FGF19 promoter in intestinal cells is mediated by PXR. *World J Gastroenterol.* 2007;13(31):4230-5.
[9]   Shang Q, Pan L, Saumoy M, Chiang JY, Tint GS, Salen G, Xu G. An overlapping binding site in the CYP7A1 promoter allows activation of FXR to override the stimulation by LXRalpha. *Am J Physiol Gastrointest Liver Physiol.* 2007;293(4):G817-23.
[10]   Kliewer SA, Willson TM. Regulation of xenobiotic and bile acid metabolism by the nuclear pregnane X receptor. *J Lipid Res.* 2002;43(3):359-64.
[11]   Dussault I, Yoo HD, Lin M, Wang E, Fan M, Batta AK, Salen G, Erickson SK, Forman BM. Identification of an endogenous ligand that activates pregnane X receptor-mediated sterol clearance. *Proc Natl Acad Sci U S A.* 2003;100(3):833-838.
[12]   Nishimaki-Mogami T, Une M, Fujino T, Sato Y, Tamehiro N, Kawahara Y, Shudo K, Inoue K. Identification of intermediates in the bile acid synthetic pathway as ligands for the farnesoid X receptor. *J Lipid Res.* 2004;45(8):1538-45.

[13] Ispolatov I, Yuryev A, Mazo I and Maslov S. Binding properties and evolution of homodimers in protein-protein interaction networks. *Nucleic Acids Res.* 2005;33(11):3629-3635.

[14] Yuryev A, Mulyukov Z, Kotelnikova E, Maslov S, Egorov S, Nikitin A, Daraselia N, Mazo I. Automatic pathway building in biological association networks. *BMC Bioinformatics.* 2006;7:171.

[15] Wu X, Ge H, Lemon B, Weiszmann J, Gupte J, Hawkins N, Li X, Tang J, Lindberg R, Li Y. Selective activation of FGFR4 by an FGF19 variant does not improve glucose metabolism in ob/ob mice. *Proc Natl Acad Sci U S A.* 2009;106(34):14379-84.

[16] Shih DM, Kast-Woelbern HR, Wong J, Xia YR, Edwards PA, Lusis AJ. A role for FXR and human FGF-19 in the repression of paraoxonase-1 gene expression by bile acids. *J Lipid Res.* 2006;47(2):384-92.

[17] Hubbert ML, Zhang Y, Lee FY, Edwards PA. Regulation of hepatic Insig-2 by the farnesoid X receptor. *Mol Endocrinol.* 2007;21(6):1359-69.

[18] Inagaki T, Choi M, Moschetta A, Peng L, Cummins CL, McDonald JG, Luo G, Jones SA, Goodwin B, Richardson JA, Gerard RD, Repa JJ, Mangelsdorf DJ, Kliewer SA. Fibroblast growth factor 15 functions as an enterohepatic signal to regulate bile acid homeostasis. *Cell Metab.* 2005;2(4):217-25.

[19] de Wildt SN, Kearns GL, Leeder JS, van den Anker JN. Cytochrome P450 3A: ontogeny and drug disposition. *Clin Pharmacokinet.* 1999;37(6):485-505.

[20] Evans WE, Relling MV. Pharmacogenomics: translating functional genomics into rational therapeutics. *Science.* 1999;286(5439):487-91

[21] El-Sankary W, Plant NJ, Gibson GG, Moore DJ. Regulation of the CYP3A4 gene by hydrocortisone and xenobiotics: role of the glucocorticoid and pregnane X receptors. Drug Metab Dispos. 2000;28(5):493-6

[22] Lehmann JM, McKee DD, Watson MA, Willson TM, Moore JT, Kliewer SA. The human orphan nuclear receptor PXR is activated by compounds that regulate CYP3A4 gene expression and cause drug interactions. J Clin Invest. 1998;102(5):1016-23.

[23] Feldman D. Binding of some non-steroidal anti-inflammatory drugs to glucocorticoid receptors *in vitro*. *Biochem Pharmacol.* 1978;27(8):1187-91.

[24] Furster C, Wikvall K. Identification of CYP3A4 as the major enzyme responsible for 25-hydroxylation of 5beta-cholestane-3alpha,7alpha,12alpha-triol in human liver microsomes. *Biochim Biophys Acta.* 1999;1437(1):46-52.

[25] Schaap FG, van der Gaag NA, Gouma DJ, Jansen PL. High expression of the bile salt-homeostatic hormone fibroblast growth factor 19 in the liver of patients with extrahepatic cholestasis. *Hepatology.* 2009;49(4):1228-35.

[26] Reiche M, Bachmann A, Lössner U, Blüher M, Stumvoll M, Fasshauer M. Fibroblast growth factor 19 serum levels: relation to renal function and metabolic parameters. *Horm Metab Res.* 2010;42(3):178-81.

[27] Desnoyers LR, Pai R, Ferrando RE, Hötzel K, Le T, Ross J, Carano R, D'Souza A, Qing J, Mohtashemi I, Ashkenazi A, French DM. Targeting FGF19 inhibits tumor growth in colon cancer xenograft and FGF19 transgenic hepatocellular carcinoma models. *Oncogene.* 2008;27(1):85-97.

[28] Wagner M, Zollner G, Trauner M. New molecular insights into the mechanisms of cholestasis. *J Hepatol.* 2009 Sep;51(3):565-80.

[29]   Lam P, Wang R, Ling V. Bile acid transport in sister of P-glycoprotein (ABCB11) knockout mice. *Biochemistry.* 2005 Sep 20;44(37):12598-605.

[30]   Deo AK, Bandiera SM. 3-ketocholanoic acid is the major *in vitro* human hepatic microsomal metabolite of lithocholic acid. *Drug Metab Dispos.* 2009 Sep;37(9):1938-47.

[31]   Tu H, Okamoto AY, Shan B. FXR, a bile acid receptor and biological sensor.*Trends Cardiovasc Med.* 2000 Jan;10(1):30-35.

[32]   Staudinger JL, Goodwin B, Jones SA, Hawkins-Brown D, MacKenzie KI, LaTour A, Liu Y, Klaassen CD, Brown KK, Reinhard J, Willson TM, Koller BH, Kliewer SA. The nuclear receptor PXR is a lithocholic acid sensor that protects against liver toxicity. *Proc Natl Acad Sci U S A.* 2001; 98(6):3369-74.

[33]   Xie W, Radominska-Pandya A, Shi Y, Simon CM, Nelson MC, Ong ES, Waxman DJ, Evans RM. An essential role for nuclear receptors SXR/PXR in detoxification of cholestatic bile acids. *Proc Natl Acad Sci U S A.* 2001; 98(6):3375-80

[34]   Matsubara T, Yoshinari K, Aoyama K, Sugawara M, Sekiya Y, Nagata K, Yamazoe Y. Role of vitamin D receptor in the lithocholic acid-mediated CYP3A induction *in vitro* and *in vivo. Drug Metab Dispos.* 2008 Oct;36(10):2058-63.

[35]   Ngan CH, Beglov D, Rudnitskaya AN, Kozakov D, Waxman DJ, Vajda S. The structural basis of pregnane X receptor binding promiscuity. *Biochemistry.* 2009 Dec 8;48(48):11572-81.

[36]   Li T, Chiang JY. Mechanism of rifampicin and pregnane X receptor inhibition of human cholesterol 7 alpha-hydroxylase gene transcription. *Am J Physiol Gastrointest Liver Physiol.* 2005 Jan;288(1):G74-8.

[37]   Zhao A, Lew JL, Huang L, Yu J, Zhang T, Hrywna Y, Thompson JR, de Pedro N, Blevins RA, Peláez F, Wright SD, Cui J. Human kininogen gene is transactivated by the farnesoid X receptor. *J Biol Chem.* 2003 Aug 1;278(31):28765-70.

[38]   Badhan R, Penny J. *In silico* modelling of the interaction of flavonoids with human P-glycoprotein nucleotide-binding domain. *Eur J Med Chem.* 2006;41(3):285-95.

[39]   Mares-Sámano S, Badhan R, Penny J. Identification of putative steroid-binding sites in human ABCB1 and ABCG2. *Eur J Med Chem.* 2009;44(9):3601-11.

[40]   Hoffmann U, Kroemer HK. The ABC transporters MDR1 and MRP2: multiple functions in disposition of xenobiotics and drug resistance. *Drug Metab Rev.* 2004 Oct;36(3-4):669-701.

[41]   Chiang JY. Bile acid regulation of gene expression: roles of nuclear hormone receptors. *Endocr Rev.* 2002 Aug;23(4):443-63.

[42]   Cai SY, He H, Nguyen T, Mennone A, Boyer JL. Retinoic acid represses CYP7A1 expression in human hepatocytes and HepG2 cells by FXR/RXR-dependent and independent mechanisms. *J Lipid Res.* 2010 Aug;51(8):2265-74.

[43]   Zhang Y, Castellani LW, Sinal CJ, Gonzalez FJ, Edwards PA. Peroxisome proliferator-activated receptor-gamma coactivator 1alpha (PGC-1alpha) regulates triglyceride metabolism by activation of the nuclear receptor FXR. *Genes Dev.* 2004 Jan 15;18(2):157-69.

[44]   Alvarez L, Jara P, Sánchez-Sabaté E, Hierro L, Larrauri J, Díaz MC, Camarena C, De la Vega A, Frauca E, López-Collazo E, Lapunzina P. Reduced hepatic expression of farnesoid X receptor in hereditary cholestasis associated to mutation in ATP8B1. *Hum Mol Genet.* 2004 Oct 15;13(20):2451-60.

[45]   Cho JY, Kang DW, Ma X, Ahn SH, Krausz KW, Luecke H, Idle JR, Gonzalez FJ. Metabolomics reveals a novel vitamin E metabolite and attenuated vitamin E metabolism upon PXR activation. *J Lipid Res.* 2009 May;50(5):924-37.

[46]   Aller SG, Yu J, Ward A, Weng Y, Chittaboina S, Zhuo R, Harrell PM, Trinh YT, Zhang Q, Urbatsch IL, Chang G. Structure of P-glycoprotein reveals a molecular basis for poly-specific drug binding. *Science*. 2009 Mar 27;323(5922):1718-22.

[47]   Khandelwal, A., Krasowski, M. D., Reschly, E. J., Sinz, M. W.,Swaan, P. W., and Ekins, S. (2008) Machine learning methods and docking for predicting human pregnane X receptor activation. *Chem Res Toxicol*. 21, 1457–1467.

[48]   Ohkura K, Kawaguchi Y, Watanabe Y, Masubuchi Y, Shinohara Y, Hori H. Flexible structure of cytochrome P450: promiscuity of ligand binding in the CYP3A4 heme pocket. *Anticancer Res*. 2009 Mar;29(3):935-42.

[49]   Mo SL, Zhou ZW, Yang LP, Wei MQ, Zhou SF. New insights into the structural features and functional relevance of human cytochrome P450 2C9. Part I. *Curr Drug Metab*. 2009 Dec;10(10):1075-126.

[50]   Ansede JH, Smith WR, Perry CH, St Claire RL 3rd, Brouwer KR. An *in vitro* assay to assess transporter-based cholestatic hepatotoxicity using sandwich-cultured rat hepatocytes. *Drug Metab Dispos*. 2010 Feb;38(2):276-80.

SUPPLEMENTARY MATERIALS

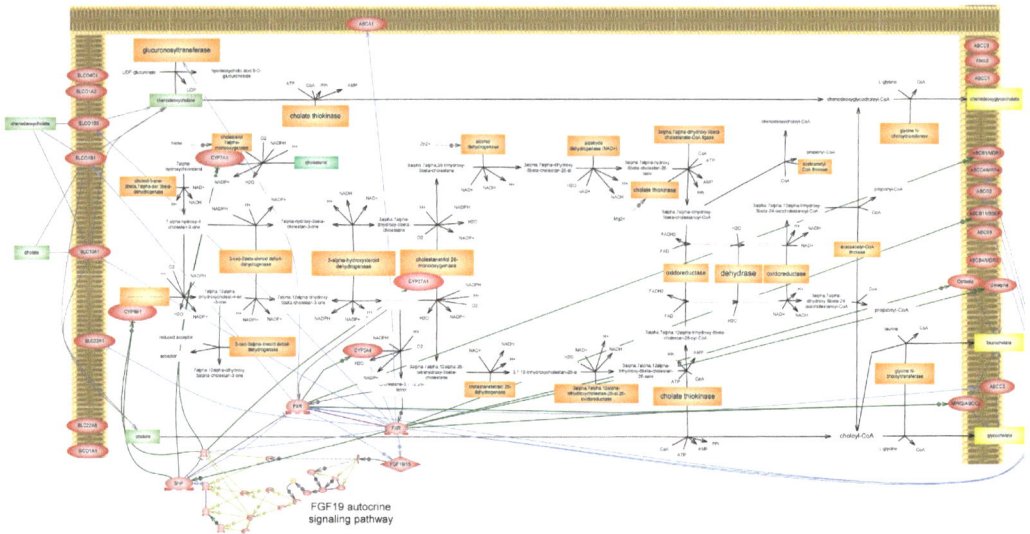

Supplementary Figure 1: The cholestasis model expanded with neutral bile acid biosynthesis pathway and reconstructed FGF19 signaling. Zoom in this page to see model details.

Supplementary Table 1: Targets of cholestasis-inducing drugs in expanded cholestasis model.

Name	Entity Type	# drugs regulating target	# drugs physically interacting with target	Functional role
FGF19 signaling pathway	Pathway	44	17	FGF19 pathway
CYP3A4	Protein	28	20	BA synthesis
ABCB1/MDR1	Protein	27	20	BA secretion
SLCO1B1/OATP2	Protein	12	12	BA import
17beta-hydroxysteroid UDP-glucuronosyltransferase	Functional Class	10	2	BA conjugation
ABCB11/BSEP	Protein	9	3	BA secretion
ABCA1	Protein	8	2	cholesterol efflux
ABCC2	Protein	7	1	BA secretion
ABCG2	Protein	7	5	BA secretion
alcohol dehydrogenase	Functional Class	7	2	BA synthesis
SLC22A8	Protein	7	6	BA import
SLCO1B3/OATP4	Protein	7	5	BA import
ABCC3	Protein	5	1	BA secretion

Slco1a1	Protein	5	3	BA import
3beta-hydroxy-delta5-steroid dehydrogenase	Functional Class	4	1	BA synthesis
ABCB4	Protein	4	0	BA secretion
ABCC1	Protein	4	1	BA secretion
ABCC4	Protein	4	0	BA secretion
CYP7A1	Protein	4	0	BA synthesis
SLCO1A2	Protein	4	4	BA import
aldehyde dehydrogenase (NAD+)	Functional Class	3	1	BA synthesis
CYP27A1	Protein	3	1	BA synthesis
SLC10A1	Protein	3	1	BA import
SLC22A1	Protein	3	3	BA import
delta4-3-oxosteroid 5beta-reductase	Functional Class	2	2	BA synthesis
cholic thiokinase	Functional Class	1	1	BA synthesis
OSTalpha	Protein	1	0	BA secretion
OSTBETA	Protein	1	0	BA secretion
SLCO4C1	Protein	1	1	BA import

Supplementary Table 2: Known cholestasis biomarkers sorted by specificity. Specificity was measured as number of disease conditions in which the biomarker level is changed. Biomarker specificity was determined as described in the *Methods* section.

Name	Description	Specificity
ADAMTS 13	ADAM metallopeptidase with thrombospondin type 1 motif, 13	10
ANPEP	alanyl (membrane) aminopeptidase	11
PENK	proenkephalin	18
CCK	cholecystokinin	28
IL4	interleukin 4	75
ADIPOQ	adiponectin, C1Q and collagen domain containing	79
PRL	prolactin	117
IL1B	interleukin 1, beta	163
MMP2	matrix metallopeptidase 2 (gelatinase A, 72kDa gelatinase, 72kDa type IV collagenase)	180
EDN1	endothelin 1	206
TGFB1	transforming growth factor, beta 1	252
TNF	tumor necrosis factor (TNF superfamily, member 2)	413

Supplementary Table 3: Proteins genetically linked to hereditary cholestasis. PMID is a PubMed identifier.

Gene	Example statement	Supporting PMID	# of References	Function
ABCB4	We show that a missense mutation in ABCB4 is a cause for ductopenic Cholestatic liver disease in adulthood., In summary, our data support a role of ABCB11 and ABCB4 mutations and polymorphisms in drug-induced cholestasis., Moreover, MDR3 mutations predispose to cholestasis of pregnancy and drug-induced cholestasis., We concluded that mutations in MDR3 accounted for approximately 2% (1/47) of infantile onset chronic cholestasis in Taiwan., These data suggest that mutation in the canalicular mdr2 is an important factor during the development of progressive familial cholestasis. <more data available.>	18781607, 17264802, 17295178, 11420418, 12967592, 17241866, 17414143, 18482588, 12381474, 12206920	24	Bile acid excretion
ABCB11	In summary, our data support a role of ABCB11 and ABCB4 mutations and polymorphisms in drug-induced cholestasis., Bile salt export pump (BSEP) deficiency is a hereditary cholestatic syndrome that results from mutations in the ABCB11 (ATP-binding cassette B11) gene., Besides rare mutations that have been linked to drug-induced cholestasis, the common p.V444A polymorphism of BSEP has been identified as a potential risk factor., In humans, mutations in the BSEP gene are associated with a very low level of bile acid secretion and severe cholestasis. <more data available.>	17264802, 20583290, 20422497, 16156672, 17241866, 18829893, 15317749, 15077010, 16763017, 18692205	19	Bile acid excretion
VPS33B	The phenotypes associated with VPS33B mutation may include incomplete Arthrogryposis-renal dysfunction-cholestasis., Arthrogryposis, Renal dysfunction and Cholestasis syndrome is a multi-system autosomal recessive disorder caused by germline mutations in VPS33B., Germline VPS33B mutations were detected in 28/35 families (48/62 individuals) with Arthrogryposis, renal dysfunction and cholestasis syndrome., Arthrogryposis, renal dysfunction and cholestasis syndrome (MIM 208085) is an autosomal recessive multisystem disorder	16492441, 18853461, 16896922, 16896922, 18972129, 16492441, 15768832, 18347289, 18853461, 16896922	15	Vesicle mediated protein sorting

	that may be associated with germline VPS33B mutations. <more data available.>			
ATP8B1	Mutations in the FIC1 gene cause relapsing or permanent cholestasis., Progressive familial intrahepatic cholestasis type 1 (PFIC1) is a specific form of genetic cholestasis caused by functional defects in FIC1/ATP8B1., Recurrent intrahepatic cholestasis (previously benign recurrent cholestasis), is also linked to specific mutations in the FIC1 gene., Testing the chip on subjects with cholestatic syndromes identified disease-causing mutations in SERPINA1, JAG1, ATP8B1, ABCB11, or ABCB4., Mutations in ATP8B1 (FIC1) underlie cases of cholestatic disease, ranging from chronic and progressive (progressive familial intrahepatic cholestasis) to intermittent (benign recurrent intrahepatic cholestasis). <more data available.>	14708891, 19381753, 10975791, 17241866, 20126555, 11093741, 15888793, 11093741, 15239083, 15317749 <more data available.>	12	aminophospholi pid translocase
ESR1	The present data indicate that polymorphism of the ERalpha and MDR3 genes 1712delT mutation are unlikely to play any significant role in obstetric cholestasis in affected Finnish women., To investigate the contribution of the estrogen receptor alpha (ERalpha) polymorphism in the development of obstetric cholestasis and to determine whether multidrug resistance 3 (MDR3) gene 1712delT mutation detected in French patients is also present in Finnish women with obstetric cholestasis., Multidrug resistance 3 gene mutation 1712delT and estrogen receptor alpha gene polymorphisms in Finnish women with obstetric cholestasis. <more data available.>	12381474, 12206920, 12381474, 12206920, 12381474	5	Estrogen receptor
KRAS	Among the 26 patients with normal or non-contributive (due to cholestasis) serum carbohydrate antigen 19.9 levels, 14 (54%) had KRAS2 mutations., Real-time PCR with a cysteine-specific (TGT) sensor probe can rapidly detect K-ras gene mutations in bile and diagnose malignant biliary obstruction with high specificity., In addition, the addition of KRAS mutation detection may improve the diagnostic accuracy of brush cytology done in extrahepatic biliary obstruction	12189555, 14718395, 16648548	3	FGF19 signaling

	(14).			
SERPIN A1	Testing the chip on subjects with cholestatic syndromes identified disease-causing mutations in SERPINA1, JAG1, ATP8B1, ABCB11, or ABCB4., We believe that genetic alterations of alpha-1 antitrypsin and P-glycoprotein, either alone or in association with known pathogenetic mechanisms, may explain the onset of danazol induced cholestasis and justify the difference in its varying duration and intensity., Apoptosis, a prominent form of cell death, is a cardinal feature of many acute and chronic liver diseases including that occurring during viral hepatitis, cholestasis, mutations in a1-antitrypsin, copper overload states, unhealthy alcohol use, nonalcoholic fatty liver disease, and ischemia-reperfusion liver injury (9).	17241866, 9360432, 16286505	3	alpha-1 protease inhibitor
HSD3B7	We conclude that a diverse spectrum of mutations in the HSD3B7 gene underlies this rare form of neonatal cholestasis., Thus, we found mutations in the HSD3B7 gene accounting for autosomal recessive neonatal cholestasis caused by 3[beta]-hydroxy-[DELTA]5-C27-steroid dehydrogenase/isomerase deficiency., In addition, we have characterized a mutation in the C27 3ß-HSD gene from a patient with neonatal cholestasis, which confirms the central role of this enzyme in bile acid synthesis.	12679481, 20531254, 11067870	3	Bile acid synthesis
CYP7B1	Mutation in CYP7B1 caused neonatal cholestasis., A mutation in the CYP7B1 gene causes severe neonatal cholestasis in a child (27)., These authors described an infant with a mutation in the CYP7B1 gene who suffered from severe cholestasis.	11470525, 12576301, 11971943	3	Bile acid synthesis (alternative pathway)
JAG1	Testing the chip on subjects with cholestatic syndromes identified disease-causing mutations in SERPINA1, JAG1, ATP8B1, ABCB11, or ABCB4., Diagnosis of Alagille syndrome, a condition that should be suspected in all patients with unexplained cholestasis, will thus be confirmed by genetic analysis for mutations of JAG1.	17241866, 10975791	2	Notch 1 ligand, tissue development
ABCC2	Mutation of BSEP and MRP2 genes in humans associated with cholestatic	15644430	1	Bile acid excretion

	disorders (Trauner *et al.,* 1998[Go]; Kullak-Ublick *et al.,* 2003[Go]).			
MDR	Mutations in ATP-binding cassette transporters cause or contribute to many different Mendelian and complex disorders including adrenoleukodystrophy, cystic fibrosis, retinal degeneration, hypercholesterolemia, and cholestasis.	16124856	1	Bile acid excretion
ABCB1	We believe that genetic alterations of alpha-1 antitrypsin and P-glycoprotein, either alone or in association with known pathogenetic mechanisms, may explain the onset of danazol induced cholestasis and justify the difference in its varying duration and intensity.	9360432	1	Bile acid excretion
OPRM1	The possibility of protection from pruritus associated with A118G supports the study of genetic polymorphisms of the OPRM1 gene in patients with cholestasis.	18709298	1	opioid receptor
NOTCH2	Glomerular mesangiolipidosis, the glomerular lesion in Alagille syndrome (OMIM# 118450), a multisystem disorder featuring severe cholestasis caused by JAG1 or NOTCH2 mutations,37 is characterized by lipid deposits within mesangial cells and matrix38 and bears little resemblance to glomerular lesions demonstrated in Notch2del1/+;Jag1+/- mice, which exhibit extrarenal phenotypes observed in humans with Alagille syndrome.39 Because glomerular mesangiolipidosis is reported in conditions of abnormal lipid metabolism,38 glomerulopathy in humans with Alagille syndrome is more likely to be an effect of cholestasis rather than a primary glomerular defect in Notch signaling.	18337488	1	tissue development
ACE	The present data indicate that the DD genotype is a genetic marker associated with an elevated risk of obstetric cholestasis, but this polymorphism of the angiotensin-converting enzyme gene is unlikely to play any significant role in preeclampsia.	11568784	1	blood pressure
SLC25A1 3	To explore the major etiological features of cholestatic liver disease in children, and to investigate the molecular epidemiological distribution of	19951499	1	mitochondrial aspartate-glutamate antiporter

	SLC25A13 mutations in cholestatic liver disease.			
SLC10A1	Our current findings would suggest, however, that patients of Asian or African descent who present with hypercholanemia of unexplained etiology or ill defined cholestatic liver disease should be further investigated for mutations in NTCP.	14660639	1	bile acid import
UGT1A1	Drug-induced acute cholestatic liver damage in a patient with mutation of UGT1A1.	17607296	1	Bile acid conjugation
HNF1B	In this report, we found a novel missense mutation in the HNF-1 beta gene in a patient with neonatal cholestasis and liver dysfunction together with the common features of MODY5.	15001636	1	liver development
CYP27A1	Mutation in the sterol 27-hydroxylase gene associated with fatal cholestasis in infancy.	15795599	1	Bile acid synthesis

CHAPTER 4

Pathways Disturbed in Duchenne Muscular Dystrophy

Maria A. Shkrob[*], Mikhail A. Pyatnitskiy, Pavel K. Golovatenko-Abramov and Ekaterina A. Kotelnikova

Ariadne Genomics Inc., Rockville, MD, USA

Abstract: The underlying cause of Duchenne muscular dystrophy (DMD) – mutations in the dystrophin gene – is known for 25 years. Still many details are to be elucidated to reconstruct the complete picture of DMD pathogenesis explaining how the lack of dystrophin leads to the disease symptoms. Dissecting the complex disease into a set of disturbed pathways helps organizing already known facts and discovering new nodes important for disease progression.

We suggest three approaches to characterize DMD through pathways. First, we manually built DMD pathways based on the literature evidence to show how intersecting disease-specific pathways allows identification of common regulators in DMD which might be considered as potential drug targets. Second, we used algorithmically generated subnetworks and a set of curated expression targets pathways to analyze genes that change expression in DMD. Using collection of the predefined pathways or automatically generated subnetworks for data analysis reveals new nodes (*e.g.* ESRRA and SREBF1) and pathways (*e.g.* IL6 and IGF1 signaling) crucial for the disease but not yet covered in literature.

Keywords: Bioinformatics, calcium, data analysis, disease models, disease networks, disease pathways, DMD, drug target discovery, duchenne muscular dystrophy, dystrophin, dystrophin glycan complex, expression analysis, mechano-transduction, meta-analysis, mitochondria biogenesis, muscle remodeling, nitric oxide, oxidative stress, pathways, signaling.

INTRODUCTION

Building models is an important part of studying disease pathogenesis and searching for new drug targets and biomarkers. It is a way of aggregating data obtained in different labs using different approaches and different model organisms, allowing the creation of a holistic picture of a disease and avoiding the

---

Address correspondence to Maria A. Shkrob: Ariadne Genomics Inc., Rockville, USA;
E-mail: maria.shkrob@ariadnegenomics.com

Anton Yuryev and Nikolai Daraselia (Eds)

focus on a single protein or cell process involved in its progression. By comparing different processes important in disease development we can identify key biological molecules relevant for several disease mechanisms. These molecules are of interest for research planning and for selecting drug targets, as changing activity of one disease network node might change the behavior of the whole network, which in turn can lead to symptom improvement. Disease models can be used for experimental data interpretation and for proposing new network members based on their proximity in the network. Major drawbacks of using disease networks for data analysis are their complexity and incompleteness. These drawbacks justify breaking a disease network into a set of pathways or using the independent set of known pathways to simplify the analysis and extend the disease network with pathway members not yet studied in the context of the disease. In this chapter we describe our approach for building pathways for Duchenne Muscular Dystrophy and using them in further data interpretation.

Duchenne Muscular Dystrophy: Duchenne Muscular Dystrophy (DMD) is a progressive X-linked recessive muscular disorder affecting about 1:3500 newborn males. DMD is the most common form of muscular dystrophy and the most common sex-linked disease of men [1]. The first symptoms of the disease usually manifest in early childhood. The DMD progression is fast: muscle weakness becomes more and more severe. First it affects skeletal muscles, making a patient wheelchair bound by the age of 12; then it spreads to heart and lung muscles. Cardiomyopathy and respiratory complications, resulting from muscle weakness, are the most common causes of death in DMD patients [2]. The average life expectancy of these patients varies from late teens to early forties, and can be improved to a certain extent by respiratory support [3,4] and drug therapy [5].

Although the primary cause of the disease - mutations in the gene coding for the dystrophin protein [7,8,9] - has been known for 25 years, it is still not known how the lack of dystrophin leads to the disease symptoms. One reason why there is little known about this relationship is due to the complexity of DMD pathogenesis. The disease originated by mutation of one gene strikes many levels of organization: multiple functional groups of molecules, cellular processes, cell communication, and tissue regeneration and remodeling. While the majority of research is limited to the role of individual proteins, signaling cascades, or cell processes in DMD progression, the complex nature of the disease necessitates an integral and systematic approach.

Building a DMD model is challenging due to the incredible complexity of its pathogenesis. The model should include several cell types, *e.g.*, both mature and non-differentiated muscle cells, myofibers of different types, cells of immune system, that have been shown to significantly contribute to disease progression, fibroblasts promoting fibrosis, and others. Since DMD is a progressive disorder, the model should reflect different disease stages. It is well-documented that DMD early stage and DMD advanced stage are different is in many ways. The fact that the disease is progressive means that there are several internal positive feedback loops in a disease model: several hubs activate each other, along with upstream and downstream targets. All this makes tracing the information flow in this system extremely complicated and the disease model is based on the molecular network rather than on pathways. Here we describe the manual construction of several overview pathways summarizing events underlying DMD pathology that can be used in new data analysis. Pathways, based on the data mined from peer-reviewed publications, were constructed in Ariadne Pathway Studio 8 software using relations from the ResNet8 database.

RESULTS: BUILDING DMD OVERVIEW PATHWAYS

Overview of DMD Pathogenesis: The original cause of DMD results from various mutations of the dystrophin gene, causing dystrophin protein deficiency in skeletal muscle [8] and in other tissues normally expressing DMD isoforms, such as cardiac muscle, brain, retina and smooth muscle [8-11]. This deficiency triggers a series of unfavorable events in muscle, as schematically represented in Fig. **1**. A more detailed scheme showing how a lack of dystrophin affects skeletal muscle is shown in Fig. **2**.

The most effective DMD treatment, evident from Fig. **1**, should target the processes at the top of the scheme. initiating all other malignant events. Drugs helping to overcome secondary disease effects (*e.g.*, inflammation, oxidative stress, and calcium overload) do not cure the disease but only ameliorate some of its symptoms. The only cure for DMD is the restoration of dystrophin production, and there are several approaches attempting to achieve it: gene therapy using a short form of dystrophin, specific anti-sense oligos inhibiting splicing and thereby restoring the dystrophin mRNA reading frame, treatment with aminoglycoside antibiotics to stimulate skipping of nonsense mutations [12]. All these approaches are currently in development. None of them has been approved to be safe and effective in humans as yet. Multiple issues have to be overcome in order to use these approaches in humans, and before it is done, therapies targeting the consequences of dystrophin absence are still in demand.

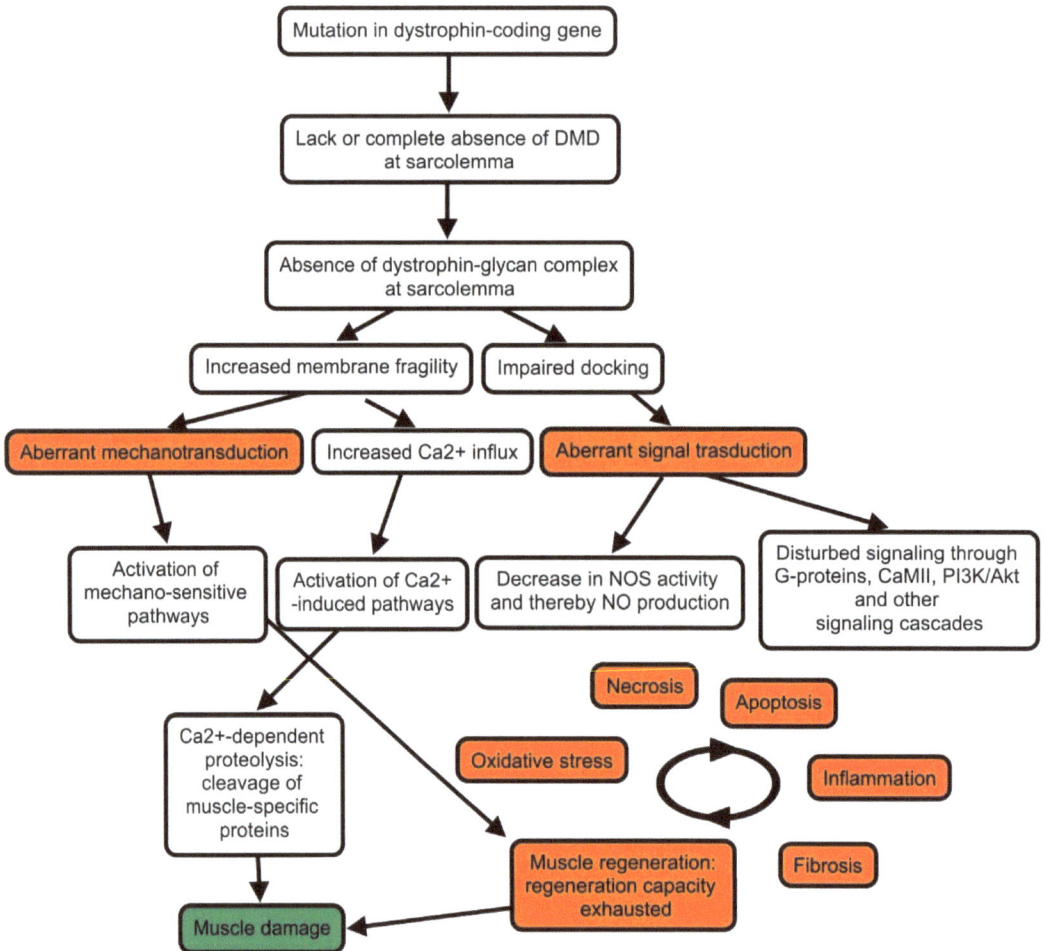

Figure 1: A schematic representation of the consequences of mutation in the dystrophin gene in muscles. In the absence of dystrophin-glycan complex, the muscle cell suffers disturbed signaling leading to cell death. Inflammation and fibrosis start at the site of injury and are constantly activated in DMD. The muscle attempts to regenerate until the regeneration capacity becomes exhausted.

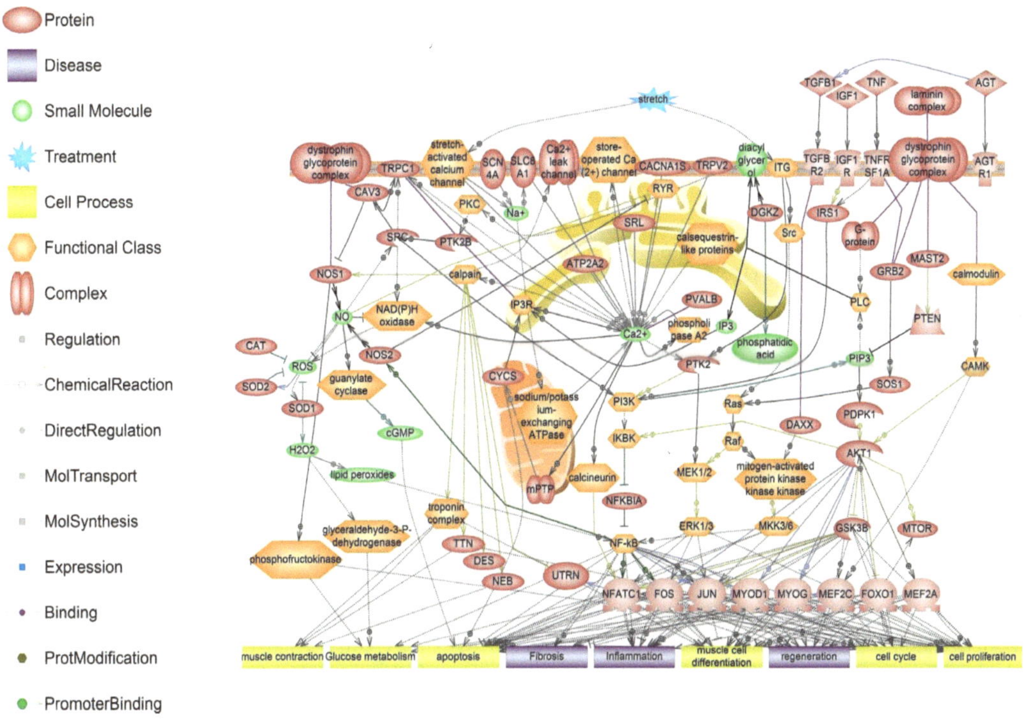

Protein

Disease

Small Molecule

Treatment

Cell Process

Functional Class

Complex

Regulation

ChemicalReaction

DirectRegulation

MolTransport

MolSynthesis

Expression

Binding

ProtModification

PromoterBinding

Figure 2: An overview scheme showing the consequences of mutations in the gene coding for dystrophin in a muscle cell. A lack of DGC causes an increased Ca^{2+} level, oxidative stress, decreased NO production and impaired signaling.

Consequences of Dystrophin Absence: The lack of dystrophin causes disintegration of dystrophin-glycan complex (DGC) in muscles. DGC spans the cellular membrane, interacting with both extracellular and cytoplasmic proteins and thereby bonding the extracellular matrix with cytoskeleton [13,14]. DGC functions in providing mechanical reinforcement to the sarcolemma and as a scaffold protein for members of intracellular signaling cascades. DGC consists of dystrophin, sarcoglycans, dystrobrevin, syntrophins, sarcospan, and dystroglycans [15,16]. Several proteins—caveolin-3 [17], nitric oxide synthase (NOS) [18,19], and laminin α2 [13,20]—are closely associated with DGC, but usually are not considered as DGC members. In the absence of dystrophin, DGC is unstable and its concentration as well as the concentration of DGC-forming and DGC-related proteins is greatly reduced at the sarcolemma [15]. Notably, mutations of at least six other genes coding for DGC members and DGC-related proteins can cause muscular diseases, characterized by similar symptoms, such as limb-girdle muscular dystrophy [21, 22] and congenital muscular dystrophy [23], making the study of consequences of DGC malfunctioning important for these other diseases as well.

Impaired Mechanotransduction: The earliest hypothesis on how the absence of DGC results in such a severe disease suggested that DGC provides mechanical reinforcement to the sarcolemma. Prior to dystrophin discovery it had been already known that dystrophic muscles suffered mechanical injuries and had disrupted plasma membranes [24]. Later, using so-called *mdx* mice that have a mutation within the dystrophin gene, thus causing DMD-like symptoms [25], it was demonstrated that the sarcolemma of dystrophic fibers had at least six times higher number of membrane ruptures [26] and that stretch caused more sarcolemma damage in dystrophic muscles [27, 28]. Increased membrane fragility not only leads to its higher permeability [26, 28], but also causes aberrant mechanotransduction [29]. Mechanical forces are very important for muscle development and maintenance. Muscle stretch influences several major signaling pathways in muscles, such as Akt, NF-κB, MAPK, calcineurin, protein kinase C, Rho, and NOS, that regulate cell growth and death [30-34].

Increased Levels of Ca^{2+}: One of the consequences of increased membrane susceptibility to stretch is increased Ca^{2+} influx. Disturbed Ca^{2+} homeostasis is considered to be one of the key factors of DMD pathology (see Fig. **3**). In dystrophic muscle, the intracellular Ca^{2+} level is increased up to two- to threefold [35, 36]. The same observation was made in *mdx* mice [37]. The exact mechanism leading to Ca^{2+} overload is not known. Several types of Ca^{2+} channels show abnormal activity in DMD patients and *mdx* mice [38-42], in which abnormal work can result in a drastic Ca^{2+} increase. Store-operated channels (SOC) serve to replenish Ca^{2+} in a case of storage depletion. Stretch-activated channels (SAC) are regulated by muscle contraction. Calcium leak channels are activated in proximity of membrane tears [43]. These channels have been defined by their activity and function. It has been recently suggested that, on the molecular level, they might be comprised of only two proteins: TRPV1 and TRPC1 [40-46]. It is known that TRPV inhibition ameliorates DMD pathology [47] as well as SAC blocking [48].

Ca^{2+} increase causes dysfunction of Ca^{2+} coupling and buffering: basal level of IP3 increases two- to threefold, density of IP3R also increases [49], activity of voltage-dependent L-type Ca^{2+} channels is also changed [50]. Ca^{2+} deregulation can cause activation of calcium-dependent proteolysis [51]. It was shown that calpain is overexpressed [52] and its activity is increased in DMD [53, 54]. This overexpression may cause devastating effects, as many of the muscle-specific structural proteins contain calpain cleavage site, *e.g.*, titin, nebulin, desmin, troponins T and I, tropomyosin, C-protein, talin, and tubulin. Calcium-dependent degradation of these proteins can lead to Z-disc abnormalities observed in dystrophic muscles [55]. Also, NOS has a calpain cleavage site [56]. Calpain inhibitors have a protective effect on muscle, delaying the onset of muscle degeneration in *mdx* mice [57, 58]. Being activated by Ca^{2+}, calpain may lead to an additional Ca^{2+} influx by affecting SOCs permeability [59], providing a positive loop mechanism of Ca^{2+} increase.

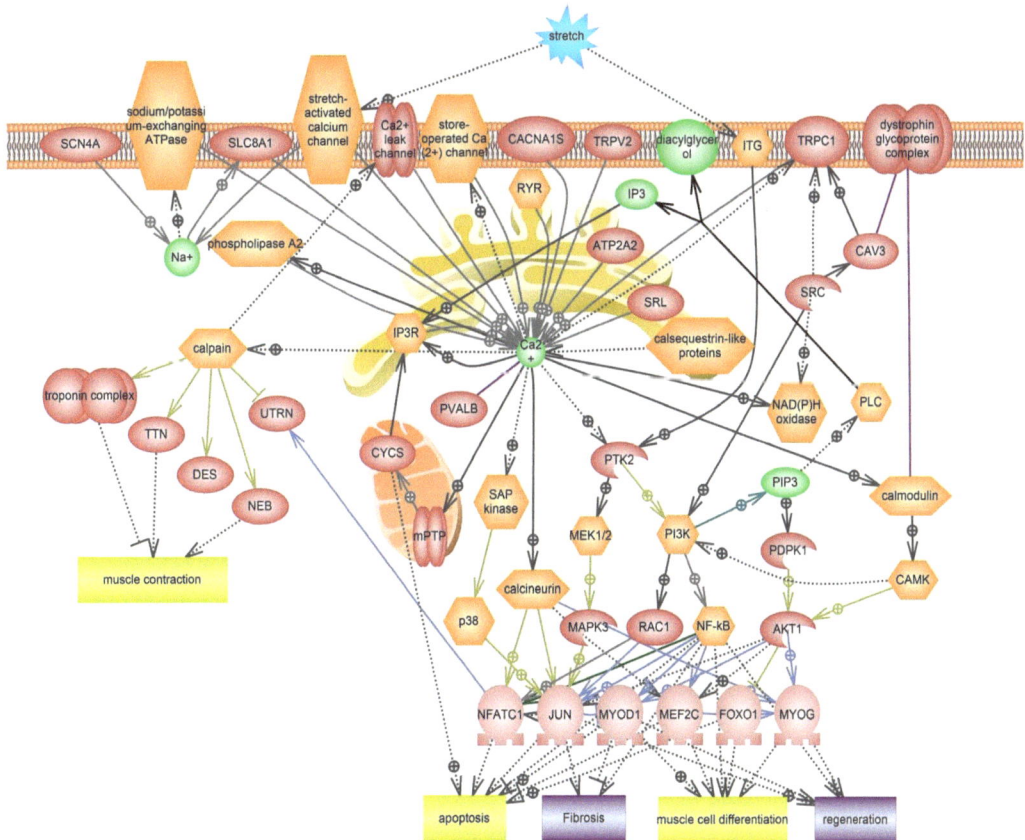

Figure 3: A scheme showing the sources of increased Ca^{2+} and the consequences of such an increase. Several types of channels, located in the sarcolemma, were shown to play role in Ca^{2+} influx in DMD. Elevated Ca^{2+} results in the activation of proteolysis, causing cleavage of some muscle-specific proteins, and activation of Ca^{2+}–dependent signaling cascades (Legend – see Fig. **2**).

Impaired Intracellular Signaling: Co-localization of several proteins involved in signal transduction depends on DGC (see Fig. **4**). Some of them, such as calmodulin kinase II [60], Grb2 [61] and NOS [18, 19] are binding to DGC itself; some are interacting with caveolin-3, a protein that is a structural component of caveolae, membrane invaginations used for scaffolding. While levels of DGC are lower in dystrophic muscle, concentration of caveolin-3 is, in contrast, higher [65]. It may contribute to the pathogenesis, as caveolin-3 serves as a transducer of a variety of signals, *e.g.*, from Src kinases [63, 64], MAPK pathway components [65], NOS [66], G-proteins [67, 68], protein kinase A and C, H-Ras [69]. They also found that caveolin-3 overexpression by itself can cause Duchenne-like symptoms in mice [70].

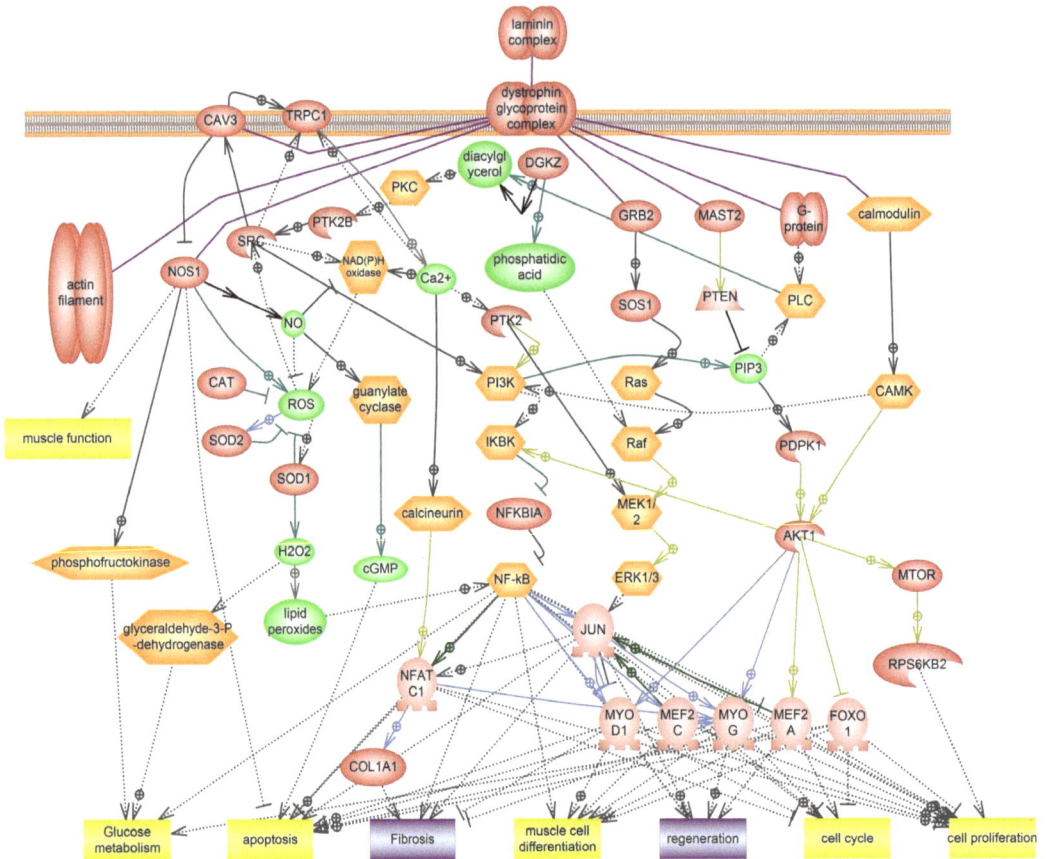

Figure 4: A scheme representing pathways disturbed by the lack of DGC. DGC and DGC-related proteins are involved in proper scaffolding of members of key signaling pathways (Legend – see Fig. **2**).

In the absence of DGC, NOS expression is decreased by more than 80% from its normal level [18,19], causing a severe reduction in NO production [71]. NO is crucial for muscle functioning, playing anti-inflammatory and cytoprotective roles as well as a signaling role important in myoblast fusion [72] and in muscle repair [73]. Activation of NO production in dystrophin-deficient muscles ameliorates some of the disease symptoms. An increase in NO can be achieved either by increasing amount of NOS [74] or by providing additional NO donors [75]. One of the newly discovered consequences of NO reduction is an abnormal pattern of histone modifications that can possibly affect DMD expression through epigenetic regulation [76].

Increased NF-κB activity was shown to play an important role in muscle wasting in dystrophin-deficient muscles [77-79]. NF-κB signaling is activated in response to oxidative stress [77], mechanical stress [30, 80], muscle unloading [81], and an elevated level of pro-inflammatory cytokines (*e.g.*, TNFα and IFNG [82]. It also promotes muscular atrophy through accelerated protein-breakdown [83]. In addition, NF-κB may promote chronic inflammation through overexpression of its downstream targets: cytokines and chemokines [84]. On the other hand, NF-κB is one of the key regulators of myogenesis and muscle repair, acting as an inhibitor of late-stage muscle differentiation through transcriptional silencing of myofibrillar genes [82, 85]. NF-κB is one of the potential drug targets for DMD management [83, 86].

Oxidative Stress: The first evidence for reactive oxygen species (ROS) involvement in DMD was obtained as early as 1971 [87]. Intracellular levels of lipid peroxidation products and thiobarbituric acid reactive products are increased in affected muscle [88, 89] As a response to an increase in ROS, anti-oxidative enzymes are overexpressed in DMD and *mdx* mice, such as glutathione peroxidase (as well as its substrate, glutathione) [90, 91], catalase, superoxide dismutase, glutathione reductase [89], glucose-6-phosphate dehydrogenase [88]. Oxidative stress precedes the onset of muscle cell death [91] and its inhibition increases strength and decreases fatigue in *mdx* mice [77].

Inflammation and Fibrosis: Inflammation, fibrosis, and impaired regeneration caused by chronic muscle damage are downstream effects of Duchenne pathology. Increased sarcolemma fragility, Ca^{2+} mishandling, and oxidative stress all lead to cell death. Biopsy specimens of dystrophic muscles contain necrotic and degenerating fibers, centrally nucleated regenerating fibers and show signs of immune cells infiltration. In the early phase of DMD, degeneration and regeneration processes are balanced, but as regenerative capacity becomes exhausted, the muscle tissue becomes replaced by adipose and connective tissue. This imbalance is extremely important for disease progression: for example, in *mdx* mice telomere activity is higher, which provides greater regenerative potential and may explain the fact that the Duchenne-like condition in *mdx* mice is much less severe than in humans [92]. Several gene expression studies reveal an inflammatory and fibrotic signature in dystrophin-deficient muscles [93-98], demonstrating increased cytokines and chemokines production, markers of invasive cells, leukocyte adhesion and diapedesis, complement system activation, and elevated expression of ECM remodeling genes. Neutrophils and macrophages infiltrate dystrophin-deficient muscle before the onset of the disease [99,100], and contribute to pathogenesis [101]. The main contributors to muscle degeneration are macrophages and T-lymphocytes infiltrating dystrophic muscle throughout all stages of the disease [99,102]. Mast cells [103], eosinophils [103], neutrophils [104], and CD4+ T-cells [99] were also shown to play a role in muscle wasting.

The key regulators involved in fibrosis promotion were defined as TGFB and CTGF [105,106].

Muscle Remodeling: Another major change in muscle affected by DMD is a change in muscle tissue recomposing. Muscles are capable of adapting their physiological, morphological and biochemical state to current demands through the process of muscle remodeling (see Fig. **5**). Each muscle consists of myofibers that differ in size, metabolism and contractile function. Slow-twitch fibers are rich in mitochondria content, have oxidative metabolism, have decreased insulin sensitivity and are more resistant to fatigue. Fast-twitch fibers are glycolytic and function in quick contractions. The transition between these two fiber types happens through gene expression [107]. The key factors of muscle remodeling are regulators of metabolism and mitochondria biogenesis, *e.g.*, PPARGC1A, AMPK, and calcineurin. It is known that DMD preferentially affects fast-twitch myofibers [108], while slow-twitch fibers show less damage. The higher survivability of slow fibers can be explained by upregulation of utrophin, a dystrophin homolog that can serve as a partial replacement for dystrophin [109]. Stimulation of fast-to-slow myofiber transition by overexpression of key regulators or using their agonists has a positive effect on muscle condition [110,111,112,113].

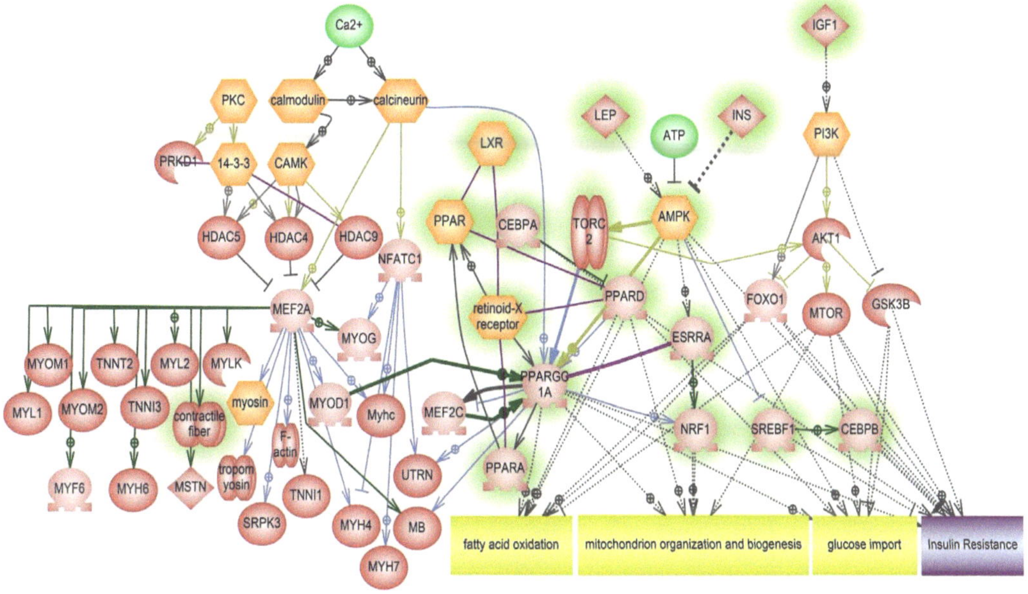

Figure 5: A scheme showing some of the key entities involved in skeletal muscle remodeling, including regulators of shifts in composition and metabolism. Regulators, derived from an analysis of the dataset (see the following section) are added to the scheme and highlighted in green (Legend – see Fig. **2**).

RESULTS: USING OVERVIEW PATHWAYS FOR SELECTING DMD TARGETS AND FOR BIOMARKER PREDICTION

An understanding of the mechanisms of DMD progression is important for searching for new therapy options. Building DMD overview pathways revealed a multifaceted process involving several major pathways and cell processes. Despite of the complexity it is important to know the entire picture, instead of studying separate regulators or partial pathways. All DMD-related processes described previously share some common regulators (*e.g.*, NOS, NF-κB, calcineurin, and Akt-PI3K pathway). This fact must be kept in mind when creating therapy approaches. On the one hand, a regulator involved in several processes is a better target (in case it has only a beneficial or only a debilitating effect on disease progression in every process). For example, there are several successful approaches to restore NO production. NO participates in almost every process affected in DMD: muscle regeneration, oxidative stress and inflammation. An increase in NO slows down the disease by decreasing inflammation, stimulation of muscle repair, and regeneration. On the other hand, we should be aware of a situation in which a treatment has favorable effects on one process while having negative effects on the others. For example, glucocorticoid prednisone, which is widely used to treat DMD patients, inhibits NF-κB signaling, has a strong anti-inflammatory action, and can decrease membrane permeability. But it also can cause shift from slow to fast myofiber content [114]. Several nodes participating in multiple processes underlying DMD pathology are already considered as drug targets for DMD. Inhibition of NFκB, stimulation of the PI3K-Akt pathway and overexpression of calcineurin in mice has showed promising results.

Targeting multiple nodes at once using combinatorial therapy enables destruction of the disease network even more effectively than mono-target therapy. For example, treatment with a compound, combining non-steroidal anti-inflammatory activity with NO-donor capability [73] or combinatorial therapy with NO-donors and glucocorticoids [75] were shown to be an effective way to ameliorate DMD symptoms.

By analyzing DMD pathways, we can infer missing information and suggest new relations and potential drug targets by proximity, *i.e.*, looking at the "neighbors" of already studied proteins on the pathway. For example, in recent years the pathway for muscle remodeling has been extensively studied in the context of DMD. Three regulators of this pathway, PPARD, PPARGC1A, and AMPK, were already shown to be important in disease progression in *mdx* mice and the golden retriever dog model of DMD and were suggested as potential drug targets. Knowing that these proteins are grouped into one part of the DMD network, we can extend it by adding their neighbors, *e.g.*, mTOR, in order to claim that they might also be of interest.

The creation of a complete but complicated disease network consisting of numerous entities, all interacting with each other in an often contradicting way, is necessary for accurate reconstruction of all events taking place in dystrophin-deficient muscle. For some tasks, however, it is useful to break this network into several pathways to simplify the picture and to focus only on major processes. The pathway approach also can be used for gene expression analysis to understand mechanisms underlying expression changes. An attempt to attach downstream expression targets to the expression regulators in the overview pathways may not provide us with a clear picture because it is difficult to trace the information flow in such a complicated network.

Splitting the disease overview pathway into sub-pathways can be also useful in the analysis of individual differences between patients so that personalized therapy approaches can be created. Building a disease mechanism from canonical signaling pathways also addresses the problem of the completeness of the disease network. The disease network built only from entities studied in the context of DMD may miss important players linked to the known studied entities found only in canonical pathways.

RESULTS: ANALYSIS OF REFERENCE EXPRESSION DATASET

Identification of Pathways Affected by DMD: We used the collection of Expression targets pathways to analyze gene expression. Expression target pathways were built based on the Ariadne Signaling Pathways collection where every pathway starts with a receptor on a surface of a cell and ends with transcription factors known to be activated by the receptor. Signaling pathways were extended from the receptor and its transcription factors, by adding ligands activating the receptor and expression targets of transcription factors. The following criteria for adding an expression target were used:

1. The target must be regulated by both the receptor and its ligand through relation type Expression regulation;

2. The target must be regulated by transcription factors through relation type Promoter binding;

3. All effect signs for a target must be consistent for all regulators. (For example, if a ligand inhibits a target, then relations ligand->target and transcription factor-> target must have opposite effect signs).

4. If the expression target meeting the requirements is a transcription factor, then its targets are also added to the Expression target pathway.

To illustrate a pathway analysis approach with expression targets pathways, we used data from [115] that contained lists of up- and downregulated genes, which were extracted from peer-reviewed publications and focused on skeletal muscle development or disease. We limited this published dataset to include only genes from DMD-related studies, which gave us a list of 2,227 genes reported to be differentially expressed from at least one publication before December 2005. The score for each gene was defined based on the direction of expression change. For every publication in which the gene was shown to be upregulated, one point was added to the score; for each publication in which the gene was shown to be downregulated one point was subtracted. Selected differentially expressed genes were imported to Pathway Studio 8 together with their scores as an Experiment. The menu option Find Pathway/Groups Enriched with Selected Entities: Expression Targets Pathways was applied. This tool returns all pathways that share at least one member from the input gene list. Found pathways are sorted by the degree of similarity to the query gene list, where similarity p-value is calculated using Fisher's exact test [116].

The top 20 pathways from our analysis are presented in Table **1**.

Table 1: The top 20 Expression targets pathways enriched with genes differentially expressed in DMD. Pathways whose contribution to DMD pathogenesis is already known are indicated in bold font. The Expanded # of Entities column shows the number of entities in the pathway, including members of Functional class entities.

Expression Targets of	Total Entities	Expanded # of Entities	Overlap	Percent Overlap	p-value
PDGF/STAT	65	65	29	44	7.41E-07
IGF1/STAT (see Fig. **6**)	81	81	33	40	1.57E-06
IGF1/MEF/MYOD/MYOG	97	124	43	34	6.57E-06
OSM-OSMR	37	37	18	48	2.55E-05
Leptin/STAT	72	72	28	38	2.99E-05
IGF1/ELK-SRF/HIF1A/MYC/SREBF	88	111	38	34	3.43E-05
IL6	94	100	35	35	4.26E-05
Insulin/STAT	108	108	37	34	4.33E-05
Insulin/CEBPA/CTNNB/FOXA/FOXO	105	121	40	33	5.31E-05

Table 1: cont....

FGF2/STAT	82	82	30	36	6.10E-05
TGFB1-TGFBR2	67	67	26	38	6.14E-05
INHBA-ACVR2/BMPR	36	36	17	47	7.02E-05
Insulin/MEF/MYOD	112	139	44	31	7.19E-05
Leptin/ELK-SRF	63	69	26	37	0.000111
Thrombopoietin/STAT	19	19	11	57	0.000143
TGFB1-TGFBR1	90	90	31	34	0.000172
TGFB3-TGFBR2	23	23	12	52	0.000267
TNF/NF-κB	139	147	44	29	0.000314
EGF/STAT	118	118	37	31	0.000362
TGFB2-TGFBR2	34	34	15	44	0.000484

Table **1** contains both pathways known to be involved in DMD pathogenesis and pathways that have not yet been studied in the context of DMD. The involvement of TGFB, PDGF, TNF and FGF2 in the DMD mechanism is well documented. This involvement can be considered as a validation of our approach.

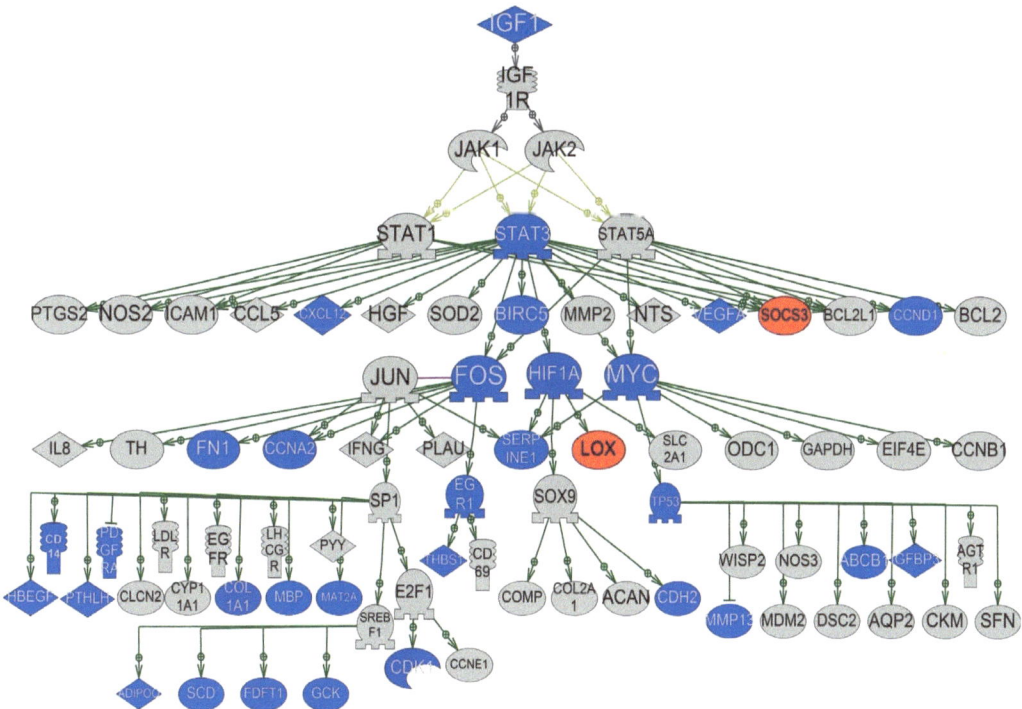

Figure 6: The IGF/STAT Signaling Expression Target Pathway. Genes showing differential expression in DMD-related peer-reviewed publications are highlighted: downregulated genes are colored blue, upregulated are colored red (Legend – see Fig. **2**).

The role of IGF-1 (insulin-like growth factor 1) pathways and IL6 (interleukin 6) pathways in DMD has not yet been well established, although both factors are known regulators of muscle function, and thereby can be of interest in DMD research. IGF-1 stimulates muscle regeneration and myogenesis through the activation of PI3K/Akt/mTOR pathway. It participates in mechanical signal transduction and withstanding of muscular atrophy [117,118]. IL6 is a myokine: a signal molecule that is produced in contracting skeletal muscle and functions in other organs of the body, aligning muscular demands with metabolic homeostasis [119]. It affects muscular regeneration and growth, also having anti-inflammatory action. This means that IL6 might be a node connecting several pathways in the DMD network inviting further research of its role in DMD. Notably, oncostatin M (OSM) is a cytokine also belonging to IL6 group, which binds the same receptor. Both IGF-1 and IL6 affect insulin sensitivity, which may explain the three insulin pathways on the list of our top 20 expression targets pathways. Decreased expression of insulin targets might be of interest, as insulin effects are smaller in slow-twitch myofibers that are less affected by DMD [107].

Identification of Major Expression Regulators in DMD: In order to find new expression regulators in DMD, we applied the Find Subnetworks Enriched with Selected Entities algorithm in Pathway Studio 8. This algorithm identifies a set of entities from the input gene list organized in sub-networks by a specific relationship. The p-value of each sub-network depends on the overlap of the sub-network with the input entities list calculated by Fisher's exact test. We used the option Expression targets, which builds sub-networks in which input genes are regulated by known transcription factors and other types of expression regulators. The p-value of such a sub-network can be considered as a measure of activity of the expression regulator in DMD. The tool was used separately for genes with positive and negative scores in order to detect both activated and repressed expression regulators. The top 20 expression regulators for both gene lists are shown in Table **2**.

We found that regulators of genes showing increased expression in DMD are mostly related to inflammatory processes and fibrotic changes. The role of most of them in DMD is already known. Regulators of genes showing decreased expression are related to metabolism regulation, mitochondria biogenesis, and muscle remodeling. Some of them have been already studied in the context of DMD (PPARGC1A, AMPK, PPARD), while others have not yet been well-studied in DMD (ESRRA, PPARA, INS, SREBF1, NRF1, retinoid-X receptor, LXR, TORC2, LEP) even though they participate in the same cell processes and pathways with regulators linked to DMD in the literature. It is worth noting that several of the regulators of under-expressed genes are under-expressed themselves (AMPK, FOXH1, IGF1, MLXIPL, NRF1, PPARGCA1, TORC2).

Table 2: The top 20 regulators of upregulated and downregulated genes. Expression regulators that have never been studied in the context of DMD are shown in boldface.

Upregulated genes	Total # of Neighbors	Overlap	Percent Overlap	p-value
IFNG	1031	191	18	4.21E-34
TGFB1	1247	192	15	1.56E-23
PDGF	410	86	20	1.33E-18
cytokine	1075	159	14	1.53E-17
Jun/Fos	703	116	16	2.89E-16
GF	616	106	17	3.49E-16
IL10	371	76	20	6.27E-16
FGF2	498	91	18	1.29E-15
TGF family	132	41	30	2.17E-15
SPI1	230	55	23	9.91E-15
CSF1	176	47	26	1.14E-14
MAPK14	473	86	18	1.15E-14
SP1	1144	158	13	1.15E-14
TNF	1370	180	13	1.31E-14
IL4	578	98	16	1.40E-14
RHOA	141	41	28	2.68E-14
AGT	553	94	16	4.42E-14
TP53	720	112	15	7.02E-14
NF-κB	966	137	14	1.31E-13
JUN	426	78	18	1.51E-13
Downregulated genes	**Total # of Neighbors**	**Overlap**	**Percent Overlap**	**p-value**
PPARGC1A	144	34	23	5.31E-13
ESRRA	58	18	30	1.80E-09
PPARA	355	48	13	1.18E-08
INS	729	77	10	3.87E-08
PPARG	489	54	11	1.29E-06

Table 2: cont....

SREBF1	157	25	15	2.30E-06
AMPK	132	22	16	4.41E-06
CEBPA	432	47	10	9.69E-06
NFYA	66	14	20	1.65E-05
NRF1	52	12	22	2.83E-05
IGF1	480	49	10	3.42E-05
TORC2	12	6	46	3.75E-05
retinoid-X receptor	278	33	11	3.90E-05
LXR	208	27	12	4.28E-05
NRIP1	31	9	28	4.59E-05
FOXH1	18	7	36	4.75E-05
PPAR	198	26	13	4.87E-05
MLXIPL	19	7	35	6.95E-05
CEBPB	351	38	10	7.84E-05
LEP	377	40	10	7.94E-05

CONCLUSION

Each of the approaches we used (building overview pathways, analyzing data using expression target pathways and search for expression regulators) have provided us with valuable insights about the disease's mechanism and potential ways of treating this condition. Overview pathways help to identify the nodes that are key to the disease's progression. Expression target pathways and searching for expression regulators give us a clue how to extend the disease's network, adding new nodes and pathways, which help us to build a more complete network of DMD.

CONFLICT OF INTEREST

Authors do not have any conflicts of interests with respect to chapter content.

ACKNOWLEDGEMENTS

The BIO-NMD grant (EC, 7th FP, proposal #241665; www.bio-nmd.eu, to Ariadne Genomics as partner) is gratefully acknowledged.

REFERENCES

[1]   Moser, H. 1984. Duchenne muscular dystrophy: pathogenetic aspects and genetic prevention. *Human Genetics* 66 (1): 17-40.

[2]   Emery, Alan E H. 2002. The muscular dystrophies. *Lancet* 359 (9307) (February 23): 687-695. doi:10.1016/S0140-6736(02)07815-7

[3]   Simonds, A, F Muntoni, S Heather, and S Fielding. 1998. Impact of nasal ventilation on survival in hypercapnic Duchenne muscular dystrophy. *Thorax* 53 (11) (November): 949-952.

[4]   Simonds, Anita K. 2006. Recent Advances in Respiratory Care for Neuromuscular Disease*. *Chest* 130 (6) (December 1): 1879 -1886. doi:10.1378/chest.130.6.1879.

[5]   Biggar, W D, V A Harris, L Eliasoph, and B Alman. 2006. Long-term benefits of deflazacort treatment for boys with Duchenne muscular dystrophy in their second decade. *Neuromuscular Disorders: NMD* 16 (4) (April): 249-255. doi:10.1016/j.nmd.2006.01.010.

[6]   Monaco, A P, R L Neve, C Colletti-Feener, C J Bertelson, D M Kurnit, and L M Kunkel. 1986. Isolation of candidate cDNAs for portions of the Duchenne muscular dystrophy gene. *Nature* 323 (6089) (October 16): 646-650. doi:10.1038/323646a0.

[7]   Koenig, M, E P Hoffman, C J Bertelson, A P Monaco, C Feener, and L M Kunkel. 1987. Complete cloning of the Duchenne muscular dystrophy (DMD) cDNA and preliminary genomic organization of the DMD gene in normal and affected individuals. *Cell* 50 (3) (July 31): 509-517.

[8]   Hoffman, EP, RH Brown, and LM Kunkel. 1987. Dystrophin: the protein product of the Duchenne muscular dystrophy locus. *Cell* 51 (6) (December 24): 919-928.

[9]   Nudel, U, D Zuk, P Einat, E Zeelon, Z Levy, S Neuman, and D Yaffe. 1989. Duchenne muscular dystrophy gene product is not identical in muscle and brain. *Nature* 337 (6202) (January 5): 76-78. doi:10.1038/337076a0.

[10]   Klamut, H J, S B Gangopadhyay, R G Worton, and P N Ray. 1990. Molecular and functional analysis of the muscle-specific promoter region of the Duchenne muscular dystrophy gene. *Mol. Cell. Biol.* 10 (1) (January 1): 193-205.

[11]   Zhao, J, M Uchino, K Yoshioka, M Miyatake, and T Miike. 1991. Dystrophin in control and mdx retina. *Brain & Development* 13 (2): 135-137.

[12]   Pichavant, Christophe, Annemieke Aartsma-Rus, Paula R Clemens, Kay E Davies, George Dickson, Shin'ichi Takeda, Steve D Wilton, *et al.*, 2011. Current status of pharmaceutical and genetic therapeutic approaches to treat DMD. *Molecular Therapy: The Journal of the American Society of Gene Therapy* 19 (5) (May): 830-840. doi:10.1038/mt.2011.59.

[13]   Ibraghimov-Beskrovnaya, Oxana, James M. Ervasti, Cynthia J. Leveille, Clive A. Slaughter, Suzanne W. Sernett, and Kevin P. Campbell. 1992. Primary structure of dystrophin-associated glycoproteins linking dystrophin to the extracellular matrix. *Nature* 355 (6362) (February 20): 696-702. doi:10.1038/355696a0.

[14]   Campbell, K P. 1995. Three muscular dystrophies: loss of cytoskeleton-extracellular matrix linkage. *Cell* 80 (5) (March 10): 675-679.

[15]   Ervasti, J M, K Ohlendieck, S D Kahl, M G Gaver, and K P Campbell. 1990. Deficiency of a glycoprotein component of the dystrophin complex in dystrophic muscle. *Nature* 345 (6273) (May 24): 315-319. doi:10.1038/345315a0.

[16]   Yoshida, M, and E Ozawa. 1990. Glycoprotein complex anchoring dystrophin to sarcolemma. *Journal of Biochemistry* 108 (5) (November): 748-752.

[17]   Sotgia, F, J K Lee, K Das, M Bedford, T C Petrucci, P Macioce, M Sargiacomo, *et al.*, 2000. Caveolin-3 directly interacts with the C-terminal tail of beta -dystroglycan. Identification of a central WW-like domain within caveolin family members. *The Journal of Biological Chemistry* 275 (48) (December 1): 38048-38058. doi:10.1074/jbc.M005321200.

[18]   Chang, W J, S T Iannaccone, K S Lau, B S Masters, T J McCabe, K McMillan, R C Padre, M J Spencer, J G Tidball, and J T Stull. 1996. Neuronal nitric oxide synthase and dystrophin-deficient muscular dystrophy. *Proceedings of the National Academy of Sciences of the United States of America* 93 (17): 9142-9147.

[19]   Brenman, J E, D S Chao, S H Gee, A W McGee, S E Craven, D R Santillano, Z Wu, *et al.*, 1996. Interaction of nitric oxide synthase with the postsynaptic density protein PSD-95 and alpha1-syntrophin mediated by PDZ domains. *Cell* 84 (5) (March 8): 757-767.

[20]   Klietsch, R, J M Ervasti, W Arnold, K P Campbell, and A O Jorgensen. 1993. Dystrophin-glycoprotein complex and laminin colocalize to the sarcolemma and transverse tubules of cardiac muscle. *Circulation Research* 72 (2) (February): 349-360.

[21]   Duggan, D J, J R Gorospe, M Fanin, E P Hoffman, and C Angelini. 1997. Mutations in the sarcoglycan genes in patients with myopathy. *The New England Journal of Medicine* 336 (9) (February 27): 618-624. doi:10.1056/NEJM199702273360904.

[22]   Minetti, C, F Sotgia, C Bruno, P Scartezzini, P Broda, M Bado, E Masetti, *et al.*, 1998. Mutations in the caveolin-3 gene cause autosomal dominant limb-girdle muscular dystrophy. *Nature Genetics* 18 (4) (April): 365-368. doi:10.1038/ng0498-365.

[23]   Helbling-Leclerc, A, X Zhang, H Topaloglu, C Cruaud, F Tesson, J Weissenbach, F M Tomé, K Schwartz, M Fardeau, and K Tryggvason. 1995. Mutations in the laminin alpha 2-chain gene (LAMA2) cause merosin-deficient congenital muscular dystrophy. *Nature Genetics* 11 (2) (October): 216-218. doi:10.1038/ng1095-216.

[24]   Mokri, B, and A G Engel. 1975. Duchenne dystrophy: electron microscopic findings pointing to a basic or early abnormality in the plasma membrane of the muscle fiber. *Neurology* 25 (12) (December): 1111-1120.

[25]   Bulfield, G, W G Siller, P A Wight, and K J Moore. 1984. X chromosome-linked muscular dystrophy (mdx) in the mouse. *Proceedings of the National Academy of Sciences of the United States of America* 81 (4) (February): 1189-1192.

[26]   Clarke, M S, R Khakee, and P L McNeil. 1993. Loss of cytoplasmic basic fibroblast growth factor from physiologically wounded myofibers of normal and dystrophic muscle. *Journal of Cell Science* 106 (Pt 1) (September): 121-133.

[27]   Head, S I, D A Williams, and D G Stephenson. 1992. Abnormalities in structure and function of limb skeletal muscle fibres of dystrophic mdx mice. *Proceedings. Biological Sciences / The Royal Society* 248 (1322) (May 22): 163-169. doi:10.1098/rspb.1992.0058.

[28]   Petrof, B J, J B Shrager, H H Stedman, A M Kelly, and H L Sweeney. 1993. Dystrophin protects the sarcolemma from stresses developed during muscle contraction. *Proceedings of the National Academy of Sciences of the United States of America* 90 (8) (April 15): 3710-3714.

[29]   Kumar, Ashok, Niraj Khandelwal, Rahul Malya, Michael b. Reid, and Aladin M. Boriek. 2004. Loss of dystrophin causes aberrant mechanotransduction in skeletal muscle fibers. *FASEB J.* 18 (1) (January 1): 102-113. doi:10.1096/fj.03-0453com.

[30]   Dogra, Charu, Harish Changotra, Jon E Wergedal, and Ashok Kumar. 2006. Regulation of phosphatidylinositol 3-kinase (PI3K)/Akt and nuclear factor-kappa B signaling pathways in dystrophin-deficient skeletal muscle in response to mechanical stretch. *Journal of Cellular Physiology* 208 (3) (September): 575-585. doi:10.1002/jcp.20696.

[31]   Boppart, M D, M F Hirshman, K Sakamoto, R A Fielding, and L J Goodyear. 2001. Static stretch increases c-Jun NH2-terminal kinase activity and p38 phosphorylation in rat skeletal muscle. *American Journal of Physiology. Cell Physiology* 280 (2) (February): C352-358.

[32] Bassel-Duby, Rhonda, and Eric N Olson. 2003. Role of calcineurin in striated muscle: development, adaptation, and disease. *Biochemical and Biophysical Research Communications* 311 (4) (November 28): 1133-1141.

[33] Nobe, K, and R J Paul. 2001. Distinct pathways of Ca^{2+} sensitization in porcine coronary artery: effects of Rho-related kinase and protein kinase C inhibition on force and intracellular Ca^{2+}. *Circulation Research* 88 (12) (June 22): 1283-1290.

[34] Tidball, J G, M J Spencer, M Wehling, and E Lavergne. 1999. Nitric-oxide synthase is a mechanical signal transducer that modulates talin and vinculin expression. *The Journal of Biological Chemistry* 274 (46) (November 12): 33155-33160.

[35] Oberc, M A, and W K Engel. 1977. Ultrastructural localization of calcium in normal and abnormal skeletal muscle. *Laboratory Investigation; a Journal of Technical Methods and Pathology* 36 (6) (June): 566-577.

[36] Bodensteiner, J B, and A G Engel. 1978. Intracellular calcium accumulation in Duchenne dystrophy and other myopathies: a study of 567,000 muscle fibers in 114 biopsies. *Neurology* 28 (5) (May): 439-446.

[37] Hopf, F. W., P. R. Turner, W. F. Denetclaw, P. Reddy, and R. A. Steinhardt. 1996. A critical evaluation of resting intracellular free calcium regulation in dystrophic mdx muscle. *Am J Physiol Cell Physiol* 271 (4) (October 1): C1325-1339.

[38] Vandebrouck, C, G Duport, C Cognard, and G Raymond. 2001. Cationic channels in normal and dystrophic human myotubes. *Neuromuscular Disorders: NMD* 11 (1) (January): 72-79.

[39] Haws, C M, and J B Lansman. 1991. Developmental regulation of mechanosensitive calcium channels in skeletal muscle from normal and mdx mice. *Proceedings. Biological Sciences / The Royal Society* 245 (1314) (September 23): 173-177. doi:10.1098/rspb.1991.0105.

[40] Vandebrouck, Clarisse, Dominique Martin, Monique Colson-Van Schoor, Huguette Debaix, and Philippe Gailly. 2002. Involvement of TRPC in the abnormal calcium influx observed in dystrophic (mdx) mouse skeletal muscle fibers. *The Journal of Cell Biology* 158 (6) (September 16): 1089-1096. doi:10.1083/jcb.200203091.

[41] Fong, P Y, P R Turner, W F Denetclaw, and R A Steinhardt. 1990. Increased activity of calcium leak channels in myotubes of Duchenne human and mdx mouse origin. *Science* (New York, N.Y.) 250 (4981) (November 2): 673-676.

[42] Turner, P R, P Y Fong, W F Denetclaw, and R A Steinhardt. 1991. Increased calcium influx in dystrophic muscle. *The Journal of Cell Biology* 115 (6) (December): 1701-1712.

[43] McCarter, G C, and R A Steinhardt. 2000. Increased activity of calcium leak channels caused by proteolysis near sarcolemmal ruptures. *The Journal of Membrane Biology* 176 (2) (July 15): 169-174.

[44] Ducret, Thomas, Clarisse Vandebrouck, My Linh Cao, Jean Lebacq, and Philippe Gailly. 2006. Functional role of store-operated and stretch-activated channels in murine adult skeletal muscle fibres. *The Journal of Physiology* 575 (Pt 3) (September 15): 913-924. doi:10.1113/jphysiol.2006.115154.

[45] Beech, D J, K Muraki, and R Flemming. 2004. Non-selective cationic channels of smooth muscle and the mammalian homologues of Drosophila TRP. *The Journal of Physiology* 559 (Pt 3) (September 15): 685-706. doi:10.1113/jphysiol.2004.068734.

[46] Maroto, Rosario, Albert Raso, Thomas G. Wood, Alex Kurosky, Boris Martinac, and Owen P. Hamill. 2005. TRPC1 forms the stretch-activated cation channel in vertebrate cells. *Nat Cell Biol* 7 (2) (February): 179-185. doi:10.1038/ncb1218.

[47]   Iwata, Yuko, Yuki Katanosaka, Yuji Arai, Munekazu Shigekawa, and Shigeo Wakabayashi. 2009. Dominant-negative inhibition of Ca^{2+} influx *via* TRPV2 ameliorates muscular dystrophy in animal models. *Human Molecular Genetics* 18 (5) (March 1): 824 - 834. doi:10.1093/hmg/ddn408.

[48]   Yeung, Ella W, Nicholas P Whitehead, Thomas M Suchyna, Philip A Gottlieb, Frederick Sachs, and David G Allen. 2005. Effects of stretch-activated channel blockers on (Ca^{2+})i and muscle damage in the mdx mouse. *The Journal of Physiology* 562 (Pt 2) (January 15): 367-380. doi:10.1113/jphysiol.2004.075275.

[49]   Liberona, J L, J A Powell, S Shenoi, L Petherbridge, R Caviedes, and E Jaimovich. 1998. Differences in both inositol 1,4,5-trisphosphate mass and inositol 1,4,5-trisphosphate receptors between normal and dystrophic skeletal muscle cell lines. *Muscle & Nerve* 21 (7) (July): 902-909.

[50]   Friedrich, O, M Both, J M Gillis, J S Chamberlain, and RHA Fink. 2004. Mini-dystrophin restores L-type calcium currents in skeletal muscle of transgenic mdx mice. *The Journal of Physiology* 555 (Pt 1) (February 15): 251-265. doi:10.1113/jphysiol.2003.054213.

[51]   Turner, P R, T Westwood, C M Regen, and R A Steinhardt. 1988. Increased protein degradation results from elevated free calcium levels found in muscle from mdx mice. *Nature* 335 (6192) (October 20): 735-738. doi:10.1038/335735a0.

[52]   Combaret, L, D Taillandier, L Voisin, S E Samuels, O Boespflug-Tanguy, and D Attaix. 1996. No alteration in gene expression of components of the ubiquitin-proteasome proteolytic pathway in dystrophin-deficient muscles. *FEBS Letters* 393 (2-3) (September 16): 292-296.

[53]   Rabbani, N, L Moses, T E Anandavalli, and M P Anandaraj. 1984. Calcium-activated neutral protease from muscle and platelets of Duchenne muscular dystrophy cases. *Clinica Chimica Acta; International Journal of Clinical Chemistry* 143 (2) (November 15): 163-168.

[54]   Kumamoto, T, H Ueyama, S Watanabe, K Yoshioka, T Miike, D E Goll, M Ando, and T Tsuda. 1995. Immunohistochemical study of calpain and its endogenous inhibitor in the skeletal muscle of muscular dystrophy. *Acta Neuropathologica* 89 (5): 399-403.

[55]   Goll, Darrell E, ValeryY F Thompson, Hongqi Li, Wei Wei, and Jinyang Cong. 2003. The calpain system. *Physiological Reviews* 83 (3) (July): 731-801. doi:10.1152/physrev.00029.2002.

[56]   Lainé, R, and P R de Montellano. 1998. Neuronal nitric oxide synthase isoforms alpha and mu are closely related calpain-sensitive proteins. *Molecular Pharmacology* 54 (2) (August): 305-312.

[57]   Badalamente, M A, and A Stracher. 2000. Delay of muscle degeneration and necrosis in mdx mice by calpain inhibition. *Muscle & Nerve* 23 (1) (January): 106-111.

[58]   Spencer, Melissa J, and Ronald L Mellgren. 2002. Overexpression of a calpastatin transgene in mdx muscle reduces dystrophic pathology. *Human Molecular Genetics* 11 (21) (October 1): 2645-2655.

[59]   Turner, P R, R Schultz, B Ganguly, and R A Steinhardt. 1993. Proteolysis results in altered leak channel kinetics and elevated free calcium in mdx muscle. *The Journal of Membrane Biology* 133 (3) (May): 243-251.

[60]   Madhavan, R, L R Massom, and H W Jarrett. 1992. Calmodulin specifically binds three proteins of the dystrophin-glycoprotein complex. *Biochemical and Biophysical Research Communications* 185 (2) (June 15): 753-759.

[61]   Yang, Bin, Daniel Jung, David Motto, Jon Meyer, Gary Koretzky, and Kevin P. Campbell. 1995. SH3 Domain-mediated Interaction of Dystroglycan and Grb2. *Journal of Biological Chemistry* 270 (20) (May 19): 11711 -11714. doi:10.1074/jbc.270.20.11711.

[62]   Brenman, J E, D S Chao, H Xia, K Aldape, and D S Bredt. 1995. Nitric oxide synthase complexed with dystrophin and absent from skeletal muscle sarcolemma in Duchenne muscular dystrophy. *Cell* 82 (5) (September 8): 743-752.

[63]   Smythe, Gayle M, Joshua C Eby, Marie-Helene Disatnik, and Thomas A Rando. 2003. A caveolin-3 mutant that causes limb girdle muscular dystrophy type 1C disrupts Src localization and activity and induces apoptosis in skeletal myotubes. *Journal of Cell Science* 116 (Pt 23) (December 1): 4739-4749. doi:10.1242/jcs.00806.

[64]   Gervásio, Othon L, Nicholas P Whitehead, Ella W Yeung, William D Phillips, and David G Allen. 2008. TRPC1 binds to caveolin-3 and is regulated by Src kinase - role in Duchenne muscular dystrophy. *Journal of Cell Science* 121 (Pt 13) (July 1): 2246-2255. doi:10.1242/jcs.032003.

[65]   Das, Manika, Samarjit Das, and Dipak K Das. 2007. Caveolin and MAP kinase interaction in angiotensin II preconditioning of the myocardium. *Journal of Cellular and Molecular Medicine* 11 (4) (August): 788-797. doi:10.1111/j.1582-4934.2007.00067.x.

[66]   Venema, V J, H Ju, R Zou, and R C Venema. 1997. Interaction of neuronal nitric-oxide synthase with caveolin-3 in skeletal muscle. Identification of a novel caveolin scaffolding/inhibitory domain. *The Journal of Biological Chemistry* 272 (45) (November 7): 28187-28190.

[67]   Chun, M, U K Liyanage, M P Lisanti, and H F Lodish. 1994. Signal transduction of a G protein-coupled receptor in caveolae: colocalization of endothelin and its receptor with caveolin. *Proceedings of the National Academy of Sciences of the United States of America* 91 (24) (November 22): 11728 -11732.

[68]   Li, Shengwen, Takashi Okamoto, Miyoung Chun, Massimo Sargiacomo, James E. Casanova, Steen H. Hansen, Ikuo Nishimoto, and Michael P. Lisanti. 1995. Evidence for a Regulated Interaction between Heterotrimeric G Proteins and Caveolin. *Journal of Biological Chemistry* 270 (26) (June 30): 15693 -15701. doi:10.1074/jbc.270.26.15693.

[69]   Song, Kenneth S., Shengwen Li, Takashi Okamoto, Lawrence A. Quilliam, Massimo Sargiacomo, and Michael P. Lisanti. 1996. Co-purification and Direct Interaction of Ras with Caveolin, an Integral Membrane Protein of Caveolae Microdomains. *Journal of Biological Chemistry* 271 (16) (April 19): 9690 -9697. doi:10.1074/jbc.271.16.9690.

[70]   Repetto, Silvia, Massimo Bado, Paolo Broda, Giuseppe Lucania, Emiliana Masetti, Federica Sotgia, Ilaria Carbone, *et al.,* 1999. Increased Number of Caveolae and Caveolin-3 Overexpression in Duchenne Muscular Dystrophy. *Biochemical and Biophysical Research Communications* 261 (3) (August 11): 547-550. doi:10.1006/bbrc.1999.1055.

[71]   Wehling, M, M J Spencer, and J G Tidball. 2001. A nitric oxide synthase transgene ameliorates muscular dystrophy in mdx mice. *The Journal of Cell Biology* 155 (1) (October 1): 123-131. doi:10.1083/jcb.200105110.

[72]   Lee, K H, M Y Baek, K Y Moon, W K Song, C H Chung, D B Ha, and M S Kang. 1994. Nitric oxide as a messenger molecule for myoblast fusion. *Journal of Biological Chemistry* 269 (20) (May 20): 14371 -14374.

[73]   Anderson, J E. 2000. A role for nitric oxide in muscle repair: nitric oxide-mediated activation of muscle satellite cells. *Molecular Biology of the Cell* 11 (5) (May): 1859-1874.

[74] Brunelli, Silvia, Clara Sciorati, Giuseppe D'Antona, Anna Innocenzi, Diego Covarello, Beatriz G. Galvez, Cristiana Perrotta, *et al.*, 2007. Nitric oxide release combined with nonsteroidal antiinflammatory activity prevents muscular dystrophy pathology and enhances stem cell therapy. *Proceedings of the National Academy of Sciences of the United States of America* 104 (1) (January 2): 264-269. doi:10.1073/pnas.0608277104.

[75] Mizunoya, Wataru, Ritika Upadhaya, Frank J. Burczynski, Guqi Wang, and Judy E. Anderson. 2011. Nitric oxide donors improve prednisone effects on muscular dystrophy in the mdx mouse diaphragm. *American Journal of Physiology - Cell Physiology* 300 (5): C1065 -C1077. doi:10.1152/ajpcell.00482.2010.

[76] Colussi, Claudia, Aymone Gurtner, Jessica Rosati, Barbara Illi, Gianluca Ragone, Giulia Piaggio, Maurizio Moggio, *et al.*, 2009. Nitric oxide deficiency determines global chromatin changes in Duchenne muscular dystrophy. *The FASEB Journal: Official Publication of the Federation of American Societies for Experimental Biology* 23 (7) (July): 2131-2141. doi:10.1096/fj.08-115618.

[77] Messina, Sonia, Domenica Altavilla, M'hammed Aguennouz, Paolo Seminara, Letteria Minutoli, Maria C Monici, Alessandra Bitto, *et al.*, 2006. Lipid peroxidation inhibition blunts nuclear factor-kappaB activation, reduces skeletal muscle degeneration, and enhances muscle function in mdx mice. *The American Journal of Pathology* 168 (3) (March): 918-926.

[78] Monici, M C, M Aguennouz, A Mazzeo, C Messina, and G Vita. 2003. Activation of nuclear factor-kappaB in inflammatory myopathies and Duchenne muscular dystrophy. *Neurology* 60 (6) (March 25): 993-997.

[79] Kramer, Henning F, and Laurie J Goodyear. 2007. Exercise, MAPK, and NF-kappaB signaling in skeletal muscle. *Journal of Applied Physiology (Bethesda, Md.: 1985)* 103 (1) (July): 388-395. doi:10.1152/japplphysiol.00085.2007.

[80] Kumar, Ashok, and Aladin M Boriek. 2003. Mechanical stress activates the nuclear factor-kappaB pathway in skeletal muscle fibers: a possible role in Duchenne muscular dystrophy. *The FASEB Journal: Official Publication of the Federation of American Societies for Experimental Biology* 17 (3) (March): 386-396. doi:10.1096/fj.02-0542com.

[81] Hunter, R Bridge, EricJ Stevenson, Alan Koncarevic, Heather Mitchell-Felton, David A Essig, and Susan C Kandarian. 2002. Activation of an alternative NF-kappaB pathway in skeletal muscle during disuse atrophy. *The FASEB Journal: Official Publication of the Federation of American Societies for Experimental Biology* 16 (6) (April): 529-538.

[82] Guttridge, D C, M W Mayo, L V Madrid, C Y Wang, and A S Baldwin. 2000. NF-kappaB-induced loss of MyoD messenger RNA: possible role in muscle decay and cachexia. *Science* (New York, N.Y.) 289 (5488) (September 29): 2363-2366.

[83] Cai, Dongsheng, J Daniel Frantz, Nicholas E Tawa, Peter A Melendez, Byung-Chul Oh, Hart G W Lidov, Per-Olof Hasselgren, *et al.*, 2004. IKKbeta/NF-kappaB activation causes severe muscle wasting in mice. *Cell* 119 (2) (October 15): 285-298. doi:10.1016/j.cell.2004.09.027.

[84] Acharyya, Swarnali, S Armando Villalta, Nadine Bakkar, Tepmanas Bupha-Intr, Paul M L Janssen, Micheal Carathers, Zhi-Wei Li, *et al.*, 2007. Interplay of IKK/NF-kappaB signaling in macrophages and myofibers promotes muscle degeneration in Duchenne muscular dystrophy. *The Journal of Clinical Investigation* 117 (4) (April): 889-901. doi:10.1172/JCI30556.

[85]   Wang, Huating, Erin Hertlein, Nadine Bakkar, Hao Sun, Swarnali Acharyya, Jingxin Wang, Micheal Carathers, Ramana Davuluri, and Denis C Guttridge. 2007. NF-kappaB regulation of YY1 inhibits skeletal myogenesis through transcriptional silencing of myofibrillar genes. *Molecular and Cellular Biology* 27 (12) (June): 4374-4387. doi:10.1128/MCB.02020-06.

[86]   Hnia, Karim, Jérôme Gayraud, Gérald Hugon, Michèle Ramonatxo, Sabine De La Porte, Stefan Matecki, and Dominique Mornet. 2008. l-Arginine Decreases Inflammation and Modulates the Nuclear Factor-κB/Matrix Metalloproteinase Cascade in Mdx Muscle Fibers. *The American Journal of Pathology* 172 (6) (June): 1509-1519. doi:10.2353/ajpath.2008.071009.

[87]   Mendell, J. R., W. King Engel, and E. C. Derrer. 1971. Duchenne Muscular Dystrophy: Functional ischemia reproduces its characteristic lesions. *Science* 172 (3988) (June 11): 1143-1145. doi:10.1126/science.172.3988.1143.

[88]   Ragusa, R J, C K Chow, and J D Porter. 1997. Oxidative stress as a potential pathogenic mechanism in an animal model of Duchenne muscular dystrophy. *Neuromuscular Disorders: NMD* 7 (6-7) (September): 379-386.

[89]   Kar, N C, and C M Pearson. 1979. Catalase, superoxide dismutase, glutathione reductase and thiobarbituric acid-reactive products in normal and dystrophic human muscle. *Clinica Chimica Acta; International Journal of Clinical Chemistry* 94 (3) (June 15): 277-280.

[90]   Austin, L, M de Niese, A McGregor, H Arthur, A Gurusinghe, and M K Gould. 1992. Potential oxyradical damage and energy status in individual muscle fibres from degenerating muscle diseases. *Neuromuscular Disorders: NMD* 2 (1): 27-33.

[91]   Disatnik, M H, J Dhawan, Y Yu, M F Beal, M M Whirl, A A Franco, and T A Rando. 1998. Evidence of oxidative stress in mdx mouse muscle: studies of the pre-necrotic state. *Journal of the Neurological Sciences* 161 (1) (November 26): 77-84.

[92]   Sacco, Alessandra, Foteini Mourkioti, Rose Tran, Jinkuk Choi, Michael Llewellyn, Peggy Kraft, Marina Shkreli, *et al.*, 2010. Short telomeres and stem cell exhaustion model Duchenne muscular dystrophy in mdx/mTR mice. *Cell* 143 (7) (December 23): 1059-1071. doi:10.1016/j.cell.2010.11.039.

[93]   Evans, Nicholas P, Sarah A Misyak, John L Robertson, Josep Bassaganya-Riera, and Robert W Grange. 2009. Dysregulated intracellular signaling and inflammatory gene expression during initial disease onset in Duchenne muscular dystrophy. *American Journal of Physical Medicine & Rehabilitation / Association of Academic Physiatrists* 88 (6) (June): 502-522. doi:10.1097/PHM.0b013e3181a5a24f.

[94]   Haslett, Judith N, Despina Sanoudou, Alvin T Kho, Richard R Bennett, Steven A Greenberg, Isaac S Kohane, Alan H Beggs, and Louis M Kunkel. 2002. Gene expression comparison of biopsies from Duchenne muscular dystrophy (DMD) and normal skeletal muscle. *Proceedings of the National Academy of Sciences of the United States of America* 99 (23) (November 12): 15000-15005. doi:10.1073/pnas.192571199.

[95]   Haslett, Judith N, Despina Sanoudou, Alvin T Kho, Mei Han, Richard R Bennett, Isaac S Kohane, Alan H Beggs, and Louis M Kunkel. 2003. Gene expression profiling of Duchenne muscular dystrophy skeletal muscle. *Neurogenetics* 4 (4) (August): 163-171. doi:10.1007/s10048-003-0148-x.

[96]   Wong, Brenda, Donald L Gilbert, Wynn L Walker, Isaac H Liao, Lisa Lit, Boryana Stamova, Glen Jickling, Michelle Apperson, and Frank R Sharp. 2009. Gene expression in

blood of subjects with Duchenne muscular dystrophy. *Neurogenetics* 10 (2) (April): 117-125. doi:10.1007/s10048-008-0167-8.

[97]   Timmons, James A, Ola Larsson, Eva Jansson, Helene Fischer, Thomas Gustafsson, Paul L Greenhaff, John Ridden, *et al.,* 2005. Human muscle gene expression responses to endurance training provide a novel perspective on Duchenne muscular dystrophy. *The FASEB Journal: Official Publication of the Federation of American Societies for Experimental Biology* 19 (7) (May): 750-760. doi:10.1096/fj.04-1980com.

[98]   Porter, John D, Sangeeta Khanna, Henry J Kaminski, J Sunil Rao, Anita P Merriam, Chelliah R Richmonds, Patrick Leahy, Jingjin Li, Wei Guo, and Francisco H Andrade. 2002. A chronic inflammatory response dominates the skeletal muscle molecular signature in dystrophin-deficient mdx mice. *Human Molecular Genetics* 11 (3) (February 1): 263-272.

[99]   Spencer, M J, E Montecino-Rodriguez, K Dorshkind, and J G Tidball. 2001. Helper (CD4(+)) and cytotoxic (CD8(+)) T cells promote the pathology of dystrophin-deficient muscle. *Clinical Immunology (Orlando, Fla.)* 98 (2) (February): 235-243. doi:10.1006/clim.2000.4966.

[100] Carnwath, J W, and D M Shotton. 1987. Muscular dystrophy in the mdx mouse: histopathology of the soleus and extensor digitorum longus muscles. *Journal of the Neurological Sciences* 80 (1) (August): 39-54.

[101] McDouall, R M, M J Dunn, and V Dubowitz. 1990. Nature of the mononuclear infiltrate and the mechanism of muscle damage in juvenile dermatomyositis and Duchenne muscular dystrophy. *Journal of the Neurological Sciences* 99 (2-3) (November): 199-217.

[102] Gorospe, J R, M D Tharp, J Hinckley, J N Kornegay, and E P Hoffman. 1994. A role for mast cells in the progression of Duchenne muscular dystrophy? Correlations in dystrophin-deficient humans, dogs, and mice. *Journal of the Neurological Sciences* 122 (1) (March): 44-56.

[103] Cai, B, M J Spencer, G Nakamura, L Tseng-Ong, and J G Tidball. 2000. Eosinophilia of dystrophin-deficient muscle is promoted by perforin-mediated cytotoxicity by T cell effectors. *The American Journal of Pathology* 156 (5) (May): 1789-1796.

[104] Hodgetts, Stuart, Hannah Radley, Marilyn Davies, and Miranda D Grounds. 2006. Reduced necrosis of dystrophic muscle by depletion of host neutrophils, or blocking TNFalpha function with Etanercept in mdx mice. *Neuromuscular Disorders: NMD* 16 (9-10) (October): 591-602. doi:10.1016/j.nmd.2006.06.011.

[105] Chen, Y-W, K Nagaraju, M Bakay, O McIntyre, R Rawat, R Shi, and E P Hoffman. 2005. Early onset of inflammation and later involvement of TGFbeta in Duchenne muscular dystrophy. *Neurology* 65 (6) (September 27): 826-834. doi:10.1212/01.wnl.0000173836.09176.c4.

[106] Sun, Guilian, Kazuhiro Haginoya, Yanling Wu, Yoko Chiba, Tohru Nakanishi, Akira Onuma, Yuko Sato, Masaharu Takigawa, Kazuie Iinuma, and Shigeru Tsuchiya. 2008. Connective tissue growth factor is overexpressed in muscles of human muscular dystrophy. *Journal of the Neurological Sciences* 267 (1-2) (April 15): 48-56.

[107] Bassel-Duby, Rhonda, and Eric N Olson. 2006. Signaling pathways in skeletal muscle remodeling. *Annual Review of Biochemistry* 75: 19-37. doi:10.1146/annurev.biochem.75.103004.142622.

[108] Webster, C, L Silberstein, A P Hays, and H M Blau. 1988. Fast muscle fibers are preferentially affected in Duchenne muscular dystrophy. *Cell* 52 (4) (February 26): 503-513.

[109] Gramolini, Anthony O., Guy Bélanger, Jennifer M. Thompson, Joe V. Chakkalakal, and Bernard J. Jasmin. 2001. Increased expression of utrophin in a slow *vs.* a fast muscle involves posttranscriptional events. *American Journal of Physiology - Cell Physiology* 281 (4) (October 1): C1300 -C1309.

[110] Wang, Yong-Xu, Chun-Li Zhang, Ruth T Yu, Helen K Cho, Michael C Nelson, Corinne R Bayuga-Ocampo, Jungyeob Ham, Heonjoong Kang, and Ronald M Evans. 2004. Regulation of Muscle Fiber Type and Running Endurance by PPARδ. *PLoS Biology* 2 (10) (October). doi:10.1371/journal.pbio.0020294.

[111] Miura, Pedro, Joe V. Chakkalakal, Louise Boudreault, Guy Bélanger, Richard L. Hébert, Jean-Marc Renaud, and Bernard J. Jasmin. 2009. Pharmacological activation of PPARβ/δ stimulates utrophin A expression in skeletal muscle fibers and restores sarcolemmal integrity in mature mdx mice. *Human Molecular Genetics* 18 (23): 4640 -4649. doi:10.1093/hmg/ddp431.

[112] Ljubicic, Vladimir, Pedro Miura, Matthew Burt, Louise Boudreault, Shiemaa Khogali, John A. Lunde, Jean-Marc Renaud, and Bernard J. Jasmin. 2011. Chronic AMPK activation evokes the slow, oxidative myogenic program and triggers beneficial adaptations in mdx mouse skeletal muscle. *Human Molecular Genetics* 20 (17): 3478 -3493. doi:10.1093/hmg/ddr265. Hori, Yusuke S., Atsushi Kuno, Ryusuke Hosoda, Masaya Tanno, Tetsuji Miura, Kazuaki Shimamoto, and Yoshiyuki

[113] Horio. 2011. Resveratrol Ameliorates Muscular Pathology in the Dystrophic mdx Mouse, a Model for Duchenne Muscular Dystrophy. *Journal of Pharmacology and Experimental Therapeutics* 338 (3): 784 -794. doi:10.1124/jpet.111.183210.

[114] Fisher, Ivan, David Abraham, Khaled Bouri, Eric P Hoffmann, Eric P Hoffman, Francesco Muntoni, and Jennifer Morgan. 2005. Prednisolone-induced changes in dystrophic skeletal muscle. *The FASEB Journal: Official Publication of the Federation of American Societies for Experimental Biology* 19 (7) (May): 834-836. doi:10.1096/fj.04-2511fje.

[115] Jelier, Rob, Peter AC 't Hoen, Ellen Sterrenburg, Johan T den Dunnen, Gert-Jan B van Ommen, Jan A Kors, and Barend Mons. 2008. Literature-aided meta-analysis of microarray data: a compendium study on muscle development and disease. *BMC Bioinformatics* 9: 291-291. doi:10.1186/1471-2105-9-291.

[116] Fisher, Ronald. 1973. *Statistical methods for research workers.* 14th ed. New York: Hafner.

[117] Mann, Christopher J, Eusebio Perdiguero, Yacine Kharraz, Susana Aguilar, Patrizia Pessina, Antonio L Serrano, and Pura Muñoz-Cánoves. 2011. Aberrant repair and fibrosis development in skeletal muscle. *Skeletal Muscle* 1 (1): 21. doi:10.1186/2044-5040-1-21.

[118] Tidball, James G. 2005. Mechanical signal transduction in skeletal muscle growth and adaptation. *Journal of Applied Physiology (Bethesda, Md.: 1985)* 98 (5) (May): 1900-1908. doi:10.1152/japplphysiol.01178.2004.

[119] Febbraio, Mark A, and Bente Klarlund Pedersen. 2002. Muscle-derived interleukin-6: mechanisms for activation and possible biological roles. *The FASEB Journal: Official Publication of the Federation of American Societies for Experimental Biology* 16 (11) (September): 1335-1347. doi:10.1096/fj.01-0876rev.

CHAPTER 5

Mechanism of Synergistic Carcinogenesis from Hypergastrinemia and Helicobacter Infection

Anton Yuryev*

Ariadne Genomics Inc., Rockville, MD, USA

Abstract: We have built a model for gastric cancer predisposition caused by hypergastrenimia during *Helicobacter* infection. The model was built using publically available data form hypergastrenimic transgenic mice infected with *Helicobacter*. We used the model to identify potential drug targets for gastric cancer, annexin II and TRAF6, and to find prognostic biomarkers for gastric cancer predisposition due to hypergastrenimia.

Keywords: Bioinformatics, pathway analysis, biomarkers, mechanistic model, gastric cancer, stomach cancer, Helicobacter, gene expression microarray, subnetwork enrichment analysis, NF-kappaB, annexin, gastrin, Pathway Studio.

INTRODUCTION

It has been firmly established that *Helicobacter* infection increases the risk of gastric cancer as well as other gastric disorders [1,2]. Gastric cancer has one of the highest mortality rates among all cancers. It has a 20% survival rate after five years from diagnosis [3]. *Helicobacter* infects the upper gastrointestinal tract of more than 50% of the human population [4] and its infection can persist for life without treatment [5]. Twenty percent of people infected with *Helicobacter* develop gastric ulcers and 1-2% of them develop gastric cancer [6]. Chronic *Helicobacter* infection causes persistent chronic gastritis which may lead to development of gastric cancer through Correa's cascade: atrophy -> intestinal metaplasia -> dysplasia -> intestinal gastric cancer [7]. *Helicobacter* infection usually results in a mild (1.5- to 2-fold elevation) hypergastrinemia that was suggested as an additional factor contributing to the development of gastric cancer. The synergy between hypergastrinemia and *Helicobacter* infection

Address correspondence to Anton Yuryev: Ariadne Genomics Inc., Rockville, USA; E-mail: anton@ariadnegenomics.com

towards gastric cancer development was proven through transgenic mice over-expressing gastrin from insulin promoter (INS-GAS mice), which develop gastric cancer at accelerated rate (7 months *vs.* 20 months) [8] upon *Helicobacter* infection [9,10]. Other factors such as diet or gender may influence the risk of gastric cancer development after *Helicobacter* infection [9].

We used the publically available Gene Expression Omnibus (GEO) dataset measuring expression of genes in the stomaches of INS-GAS mice upon *Helicobacter felis* infection [11] to elucidate the molecular mechanism behind the hypergastrinemia and *Helicobacter* infection synergy towards gastric cancer development. We found that INS-GAS mice infected with *Helicobacter* have hyperactivated transcriptional activity of NF-κB, which eventually can shift the balance towards suppression of apoptosis and activation of cell proliferation in the classical cell proliferation-survival-apoptosis regulatory network. To provide the mechanism for synergistic activation of NF-κB by gastrin and *Helicobacter* we reconstructed gastrin->NF-κB signaling pathway since the mechanism of NF-κB activation by *Helicobacter* was established previously. Our model can be used to explain or predict additional factors contributing to gastric cancer development by activating NF-κB further beyond levels achieved due to chronic *Helicobacter* infection. The model can also be used to predict individual genetic predisposition to gastric cancer by finding SNPs activating pathway components.

RESULTS

NF-κB is a Major Transcription Factor Activated only by the Combined Action of Helicobacter Felis Infection and Hypergastrinemia: Three sets of expression regulators were calculated using sub-network enrichment analysis (SNEA) as described in the *Methods* section: 1) 134 regulators activated due to gastrin over-expression, 2) 236 regulators activated in wild-type mice upon *H. felis* infection, and 3) 281 regulators activated upon *H. felis* infection of INS-GAS mice (Supplementary Table **1**). For all conditions we found the vast majority of expression regulators (> 98%) having up-regulated targets indicating that the expression activity of the transcriptional activators was also activated rather than suppressed. To find expression regulators uniquely activated only in INS-GAS mice infected with *H. felis* we subtracted regulators activated by gastrin and regulators activated by *H. felis* infection from regulators activated upon *H. felis* infection of INS-GAS mice. We found 132 regulators activated only by the synergistic action of gastrin over-expression and *H. felis* infection and classified them into 45 secreted proteins including hormones, 36 plasma membrane receptors, 34 signal transduction and other cytoplasmic proteins, and 17 transcription factors and other nuclei DNA binding proteins (see Supplemental Table **1**).

We focused further investigation exclusively on transcription factors, assuming that *H. felis* and gastrin induce cancerogenesis directly in stomach cells by binding to plasma membrane receptors on their surface and consequently co-activating intracellular proteins. Such co-activation must ultimately lead to the activation of transcription factors inside stomach cells that shift the quiescent state of gastric cells either by inducing cell proliferation and/or inhibiting apoptosis. To prioritize the 17 found transcription factors we performed another sub-network enrichment analysis in order to find the transcription factors regulating the 132 regulators uniquely activated by *Helicobacter* synergy with hypergastrinemia. Among the 17 transcription factors identified previously we found 5 that regulated transcriptions of other regulators, *i.e.*, they are upstream of major expression regulators activated by *Helicobacter*-hypergastrinemia synergy. Three of them were components of the NF-κB complex: Protein entities NFKBIA, REL and Functional class entity NFKB1. A closer inspection of the transcription factors list upstream of 132 expression regulators stimulated by combined action of *Helicobacter* and hypergastrinemia identified additional members of the NF-κB complex: Protein entity RELB, Functional class entity NF-κB, and Protein entity IKBKE. Functional class entity NF-κB was the most significant transcription factor on the list of regulators controlling expression of 45 targets out of the 132 proteins used for input. Altogether we found 50 major expression regulators which expression is regulated by components of NF-κB-IkB-IKK complex (Fig. **1**). They, in turn, regulate 281 genes measured on the expression microarray chip out of 1,500 genes that had at least one significant expression regulator in the ResNet 8 database. Based on this analysis we concluded that NF-κB is the major transcription factor responsible for differential expression during simultaneous hypergastrinemia and *Helicobacter* infection, which is uniquely hyper-activated only when the two stimuli are applied together.

Reconstruction of Gastrin -> NF-κB Pathway: Activation of NF-κB by *Helicobacter* in murine gastric epithelial cells was reported previously [12]. Activation is most likely to happen through the TLR4 receptor that has been shown to be activated by *Helicobacter* in gastrointestinal epithelial cells [13] or through TLR2 and TLR5 receptors reported to activate NF-κB in epithelial cells after *Helicobacter* infection [14]. In support of these reports we also found TLR2 as major expression regulators uniquely activated due to the synergy between *Helicobacter* and hypergastrinemia (Supplemental Table **1**). We also found TLR-1/3/4/5/6/7/9 receptors as significant regulators of differential expression after the *H. felis* infection of wild-type non-transgene mice, while TLR-1/2/5/6/7 receptors were activated upon *H. felis* infection of INS-GAS mice. The Pathway Studio ResNet 8 database has the "TLR1/2/6 -> NF-κB signaling" and "TLR4/5/7/9 -> NF-κB signaling" canonical pathways in the Ariadne Signaling Pathways collection. Therefore, in order to build mechanistic model for synergy between gastrin and *Helicobacter* towards NF-κB activation, we needed to reconstruct the gastrin-> NF-κB signaling pathway and then combine it with the existing TLR-> NF-κB canonical signaling pathways. Reconstruction of the pathway for gastrin activation of NF-κB is detailed in the *Methods* section, and the final result is shown in Fig. **2**.

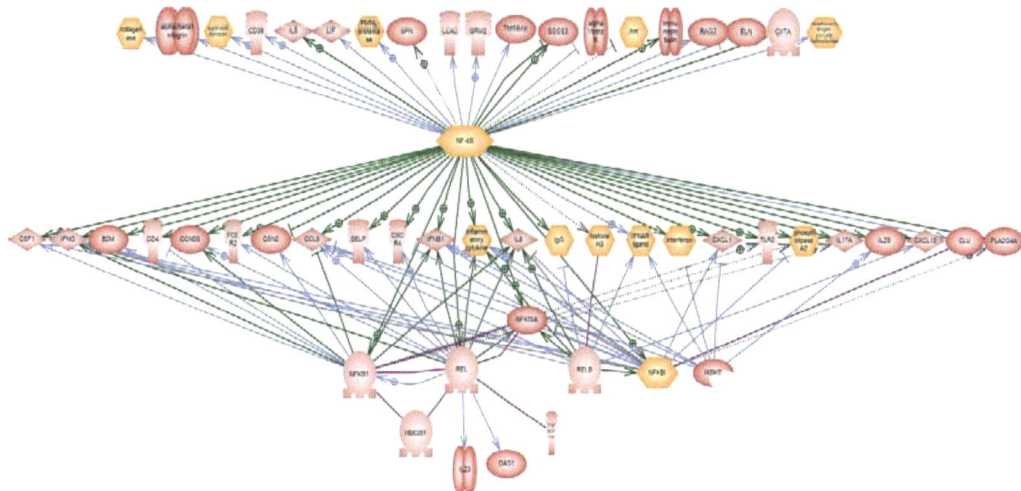

Figure 1: NF-κB and its targets, NFκBI and IkB, are known to regulate 50 out of the 132 major expression regulators identified by SNEA analysis of differential expression between INS-GAS transgene infected with *H. felis* and uninfected control.

Building a Mechanistic Model for Hypergastrinemia and Helicobacter Synergy Towards Gastric Cancer: The final model for synergistic activation of NF-κB by hypergastrinemia and *Helicobacter* infection was built by combining canonical pathways for activation of NF-κB by a toll-like receptor with a reconstructed pathway for gastrin activation of NF-κB. The result is shown in Fig. **3**.

DISCUSSION

Mechanistic Model of Synergistic Activation of NFκB by Gastrin and Helicobacter: The model developed in this study consists mostly of canonical pathways. It contains a canonical toll-like receptor pathway activating NF-κB and a portion of the gastrin canonical pathway. The model was developed based on the analysis of gene expression data which have suggested hyper-activation of NF-κB by the synergistic action of *Helicobacter* infection and hypergastrinemia. We hypothesize that prolonged activation of NF-κB significantly predisposes gastric cells to cancerogenesis due to the suppression of apoptosis and activation of cell proliferation—two well-known consequences of NF-κB activation in most tissues.

The model predicts the central role of annexin II in mediating the synergistic cross-talk between gastrin and toll-like receptor signaling. The role of annexin II in gastrin activation of NF-κB has been reported in other tissues [17, 23]. Annexin II can be secreted into an extracellular matrix where it can activate TLR4; it also can be imported into cytoplasm where it can interact with I-kappa-B kinase epsilon and TRAF6. We propose that gastrin activation of the cholecystokinin B receptor leads to the decrease of local concentration of phosphatidylinositol 4,5-biphosphate, which releases annexin II from the plasma membrane. This local release of annexin II leads to the local activation of TLR4 through the annexin II secretion and to NF-κB activation through RhoA and I-kappa-B kinase. This contributes to NF-κB activation caused by *Helicobacter* infection through toll-like receptors. The combination of prolonged hypergastrinemia with *Helicobacter* infection may promote the formation of macromolecular complexes between gastrin receptors and toll-like receptors in the plasma membrane. Such complexes can be formed through the physical interaction of gastrin with annexin II and through annexin II interaction with TRAF6, a component of the toll-like receptor signaling complex. The physical co-localization of two receptor complexes may contribute even further to permanent super-activation of NF-κB. Both annexin II and TRAF6 are druggable proteins that can be used as targets for drugs against gastric cancer. Our model suggests that inhibition of these proteins in gastric cancer will disrupt synergy between high levels of gastrin and *Helicobacter* infection, thus attenuating the risk of cancer development.

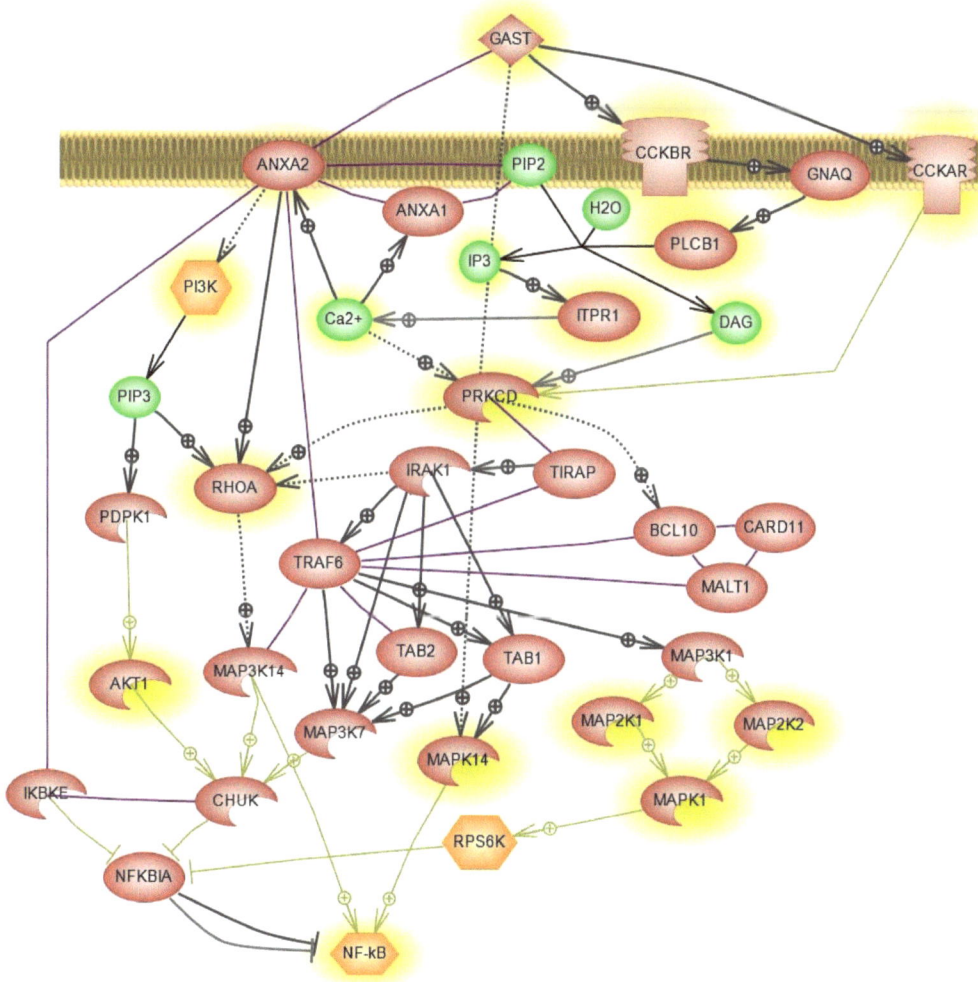

Figure 2: A reconstructed pathway of NF-κB activation by gastrin. Gastrin can activate NF-κB by two mechanisms: through PKC-delta activation by cholecystokinin receptors and through annexin II activation. Two pathways can cross-talk through phospholipase C beta that cleaves phosphatidylinositol 4,5-biphosphate and thus locally releases annexin II from the plasma membrane to allow it to bind and activate RhoA, leading to activation of NIK kinase. Annexin II itself was reported in a complex with IkB kinase epsilon, suggesting that it serves as an adaptor molecule for IkB kinase activation by gastrin. Known gastrin targets are highlighted in yellow.

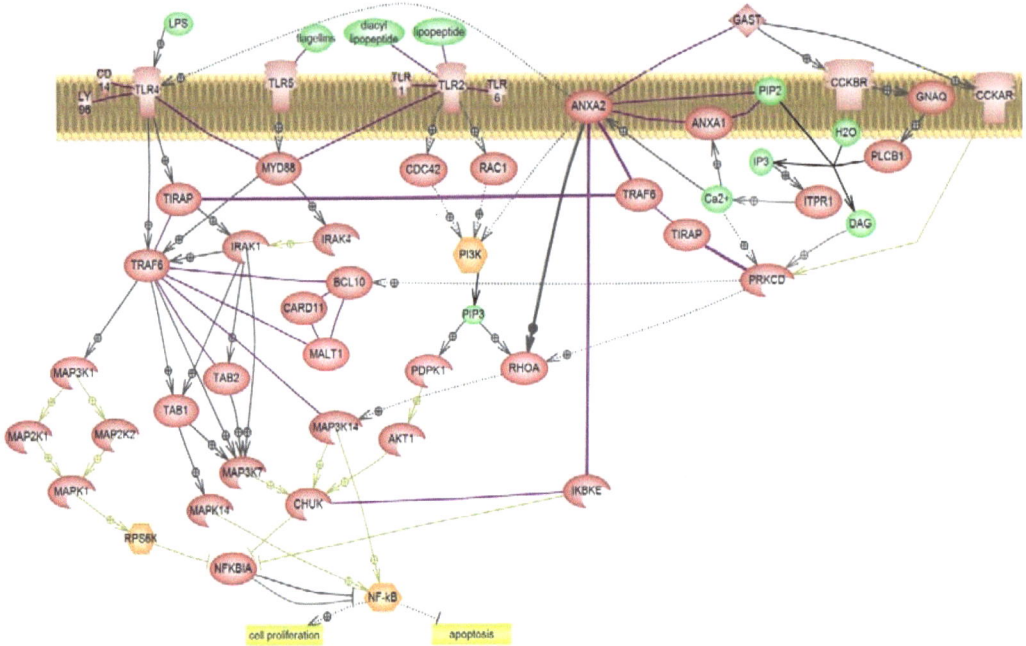

Figure 3: The final model for synergistic activation of NF-κB by hypergastrinemia and *Helicobacter* infection. The central role of annexin II (ANXA2) and PKC-delta (PRKCD) in cross-talk between gastrin and the toll-like receptor pathway is indicated by thicker lines. They reported PKC-delta regulation of RhoA [25] and BCL10 [26] as well as annexin II regulation of PI3K [27], but no mechanism could be deduced using current knowledge.

Application of the Model for Prediction of a Gastric Cancer Prognostic Biomarker: Our model suggests that biomarkers for NF-κB super-activation can be used as early indicators for patient predisposition to gastric cancer. To predict such biomarkers we have used the same gene expression dataset from GDS2103 that was used to develop the model. Functional class NF-κB has 846 expression targets in the ResNet 8 database. Therefore, we have focused only on expression targets of NFKB1 and REL – two transcription factors activated in response to *Helicobacter* infection of INS-GAS mice according to sub-network enrichment analysis of GDS2103. 123 out of 168 expression targets known for these two factors in ResNet 8 were measured in GDS2103. Clinical biomarkers must be easily assayable secreted proteins therefore we have focused on 36 extracellular proteins regulated by NF-κB. We found at least 7 candidate biomarkers which are over-expressed by NF-κB only during synergistic action of gastrin and *Helicobacter*. The biomarker selection process is shown on Fig. **4**.

To validate predicted biomarkers we used the Ariadne Disease*FX* database to find out how many biomarkers predicted in this study were previously suggested as biomarkers for gastric cancer. We found that the following biomarkers were confirmed experimentally and suggested as biomarkers for gastric cancer previously: IL1B [28], TNF [29], IL6 [28, 30], IL10 [31, 32], PLAU [33], IL2 [29], APOE [34], COL1A1 [34].

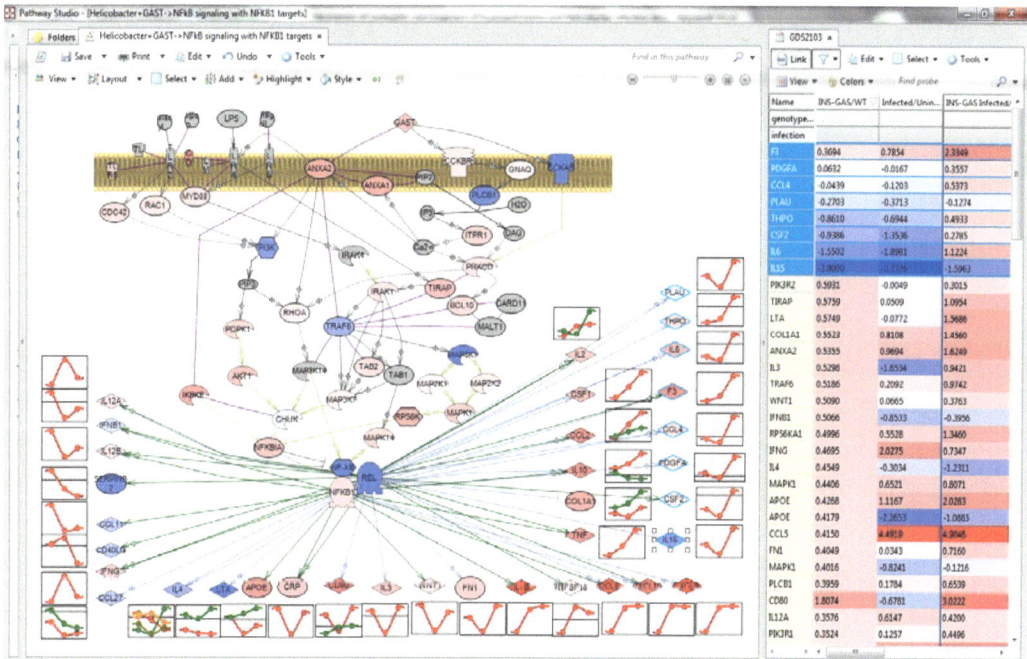

Figure 4: Candidate biomarkers indicating super-activation of NF-κB were selected as secreted proteins which expression is known to be regulated by NF-κB. Seven such proteins are over-expressed only during hypergastrenimia combined with *Helicobacter* infection. Expression profile of each secreted proteins is shown by graph next to it. Graphs that contain several lines indicate multiple probes for the same gene on the microarray chip. Each expression profile has three data points in the following order: over-expression of protein during hypergastrenimia, over-expression of protein during *Helicobacter* infection, and over-expression of protein during hypergastrenimia combined with *Helicobacter* infection. The heat map to the right shows expression of all genes in the pathway for these three data points normalized on expression in wild-type mice.

METHODS

Calculation of Differential Expression and Sub-Network Enrichment Analysis of Gene Expression Data: GDS2103 expression experiment was downloaded from gene Expression Omnibus Website in SOFT format. The file was imported into the

Pathway Studio 8 software by selecting the option "GEO Datasets (GDS in soft format)" from the Import->Experiment wizard dialog box. Differential expression was calculated between three pairs of samples: gastrin transgenic *vs.* uninfected control, *H. felis* non-transgenic control *vs.* uninfected control, *H. felis* transgene *vs.* uninfected control by selecting the "Compare groups of uncorrelated samples (2-class unpaired T-test)" option in the"Find differentially expressed Genes" dialog box.

Sub-network enrichment was performed for all three differential expression columns using the option "Expression targets" to find expression regulators upstream of the most differentially expressed genes. To find the 132 regulators activated only by the synergistic action of *H.felis* and hypergastrinemia, we added three sets of SNEA regulators into separate pathways and subtracted from regulators that were significant for the *H. felis* transgene *vs.* the uninfected control comparison regulators found in the other two comparisons. The list of 132 regulators is available in Supplementary Table **1**.

To further prioritize SNEA regulators we performed a second SNEA analysis using the menu option "Tools->find sub-network enriched with selected entities" and using the 132 regulators activated only by *H. felis* + hypergastrinemia. This option uses Fisher's exact test to find upstream regulators and does not take into account expression values of the regulators on the chip. To further select regulators that were significant in both the first and second SNEAs, we used the option "Copy gene seeds" in the SNEA result table and the option "Select->Select clipboard special" with the options "Select members of Functional class" and "Select against members of Functional Classes" selected. This option allowed us to compare SNEA results not only directly but also on the level of functional classes and their members. This approach found the NF-κB Functional class to be significant in the second SNEA, while its members REL and NFKB1 were significant in the first SNEA. Additionally, the NFKBI Functional class and its member NFKBIA were significant in first SNEA, while IKBKE was significant in the second SNEA. Both IKBKE and NFKBI are components of the canonical NF-κB activation pathway. Their significance in SNEAs provided additional support for the hypothesis that NF-κB is the principal regulator responsible for differential expression caused by the synergistic action of *H. felis* infection and hypergastrinemia.

Pathway Reconstruction for NF-κB Activation by Gastrin: We used the ResNet 8 database enhanced with three quarterly updates from Araidne Genomics and data from the PINA database [15] as a source of physical interactions for pathway reconstruction. A *Regulation* relation between gastrin and NF-κB was found using Pathway Studio Build pathway tool option "Find Direct regulation". It was supported by three publications [16, 17, 18] which reported that gastrin activates NF-κB through PKC-delta in the human gastric cancer cell line [16], and through RhoA proteins in pancreatic cancer cells [17]. The path for PKC-delta activation by gastrin through phosphatidylinositol 4,5-biphosphate degradation by PLCB1 and subsequent Ca^{2+} release was taken from the "CholecystokininR -> ELK-SRF signaling" canonical pathway available in the Ariadne Signaling pathway collection. The rest of the pathway was reconstructed based on the evidence provided in [16] which reported that NF-κB activation by gastrin requires TRAF6 and NIK and TAK1 kinases.

We closed the gap between PKC-delta and TRAF6 by finding canonical signaling pathways that contained both PKC-delta and TRAF6 proteins by using the "Find pathways/groups enriched with selected entities" command. Two pathways, "B-cell receptor->NF-κB signaling" and "CD38-> NF-κB signaling," had an identical path for TRAF6 activation by PKC: through the BCL10-CARD11-MALT1 protein complex. The references supporting PKC activation of Bcl10 confirmed that activation occurs through the PKC-delta isoform [20] in B-cells. Bcl10 was reported to be expressed in normal gastric tissues [21]; therefore, it can participate in PKC-delta signaling in response to gastrin. We also used the "Find shortest path" algorithm to find TIRAP protein as another possible way in order for PKC-delta to activate TRAF6. TIRAP is required for TRAF6 activation in the toll-like receptor signaling pathway [19] and also attracts PKC-delta in participating in toll-like receptor signaling [22]. In the toll-like receptor signaling pathway, TIRAP activates TRAF6 through IRAK1 kinase. Therefore, IRAK1 was also added to the gastrin pathway.

An additional path for gastrin activation of NF-κB through annexin II protein was found after inspecting all plasma membrane proteins that physically interact with gastrin. These proteins were found by adding all gastrin physical interactions and then using a layout by cell localization to quickly find proteins localized in the plasma membrane. Our inspection of the reference supporting gastrin - annexin II *Binding* relation revealed that annexin II is an important mediator of NF-κB activation by gastrin [23]. In the pancreatic cancer cell line, AR42J gastrin activates cell proliferation through PI3K, AKT, p38, MAPK1/2, IkappaB kinase alpha/beta, IkappaB-alpha and NF-κB component RELA. Fifteen canonical pathways in the Ariadne Signaling pathways collection contain a conserved block for NF-κB activation by PI3K *via* the PDPK1-AKT path, and four canonical pathways contain the TRAF6 – MAPK1/2 – RSK6 kinase path for NFκB activation. DDR1->NF-κB signaling pathway shows that p38 kinase can be activated by TAB1 protein to directly phosphorylate NF-κB.

The final pathway, including all conserved signaling blocks and paths reported for gastrin activation of NF-κB, is shown in Fig. **2**. It also shows that annexin II can form physical complexes with both TRAF6 and IKBE which were found in large scale protein-protein interaction mapping [24]. These interactions were found among data imported from the PINA database [15].

CONFLICT OF INTEREST

Authors do not have any conflicts of interests with respect to chapter content.

ACKNOWLEDGEMENT

None declared.

REFERENCES

[1]   Marshall BJ, Warren JR. Unidentified curved bacilli in the stomach of patients with gastritis and peptic ulceration. *Lancet* 1984;1:1311-1315.
[2]   Correa P, Houghton J. Carcinogenesis of *Helicobacter* pylori. *Gastroenterology* 2007;133:659-672.
[3]   Jemal A, Siegel R, Ward E, Hao Y, Xu J, Murray T, Thun MJ. Cancer statistics. *CA Cancer J Clin.* 2008; 58:71-9
[4]   Yamaoka Y (editor). *Helicobacter* pylori: Molecular Genetics and Cellular Biology. Caister Academic Press. 2008; ISBN 978-1-904455-31-8.
[5]   Brown LM. *Helicobacter pylori*: epidemiology and routes of transmission. *Epidemiol Rev.* 2000;22(2): 283–97
[6]   Kusters JG, van Vliet AH, Kuipers EJ. Pathogenesis of *Helicobacter pylori* infection. *Clin Microbiol Rev.* 2006;19(3): 449–9.

[7]   Correa P. Human gastric carcinogenesis: a multistep and multifactorial process - First American Cancer Society Award Lecture on Cancer Epidemiology and Prevention. *Cancer Res.* 1992 Dec 15;52(24):6735-40.

[8]   Wang TC, Dangler CA, Chen D, Goldenring JR, Koh T, Raychowdhury R, Coffey RJ, Ito S, Varro A, Dockray GJ, Fox JG: Synergistic interaction between hypergastrinemia and *Helicobacter* infection in a mouse model of gastric cancer. *Gastroenterology* 2000, 118:36-47.

[9]   Fox JG, Rogers AB, Ihrig M, Taylor NS, Whary MT, Dockray G, Varro A, Wang TC: *Helicobacter* pylori-associated gastric cancer in INS-GAS mice is gender specific. *Cancer Res.* 2003, 63:942-95.

[10]   Takaishi S, Tu S, Dubeykovskaya ZA, Whary MT, Muthupalani S, Rickman BH, Rogers AB, Lertkowit N, Varro A, Fox JG, Wang TC. Gastrin is an essential cofactor for *Helicobacter*-associated gastric corpus carcinogenesis in C57BL/6 mice. *Am J Pathol.* 2009 Jul;175(1):365-75.

[11]   Takaishi S, Wang TC. Gene expression profiling in a mouse model of *Helicobacter*-induced gastric cancer. *Cancer Sci.* 2007;98(3):284-93.

[12]   Ferrero RL, Avé P, Ndiaye D, Bambou JC, Huerre MR, Philpott DJ, Mémet S. NF-kappaB activation during acute Helicobacter pylori infection in mice. *Infect Immun.* 2008 Feb;76(2):551-61.

[13]   Su B, Ceponis PJ, Lebel S, Huynh H, Sherman PM. *Helicobacter* pylori activates Toll-like receptor 4 expression in gastrointestinal epithelial cells. *Infect Immun.* 2003 Jun;71(6):3496-502.

[14]   Smith MF Jr, Mitchell A, Li G, Ding S, Fitzmaurice AM, Ryan K, Crowe S, Goldberg JB. Toll-like receptor (TLR) 2 and TLR5, but not TLR4, are required for *Helicobacter* pylori-induced NF-kappa B activation and chemokine expression by epithelial cells. *J Biol Chem.* 2003 Aug 29;278(35):32552-60. Epub 2003 Jun 13.

[15]   Wu, J., Vallenius, T., Ovaska, K., Westermarck, J., Makela, T.P. and Hautaniemi, S. (2009) Integrated network analysis platform for protein-protein interactions, *Nature methods* 6, 75-77.

[16]   Ogasa M, Miyazaki Y, Hiraoka S, Kitamura S, Nagasawa Y, Kishida O, Miyazaki T, Kiyohara T, Shinomura Y, Matsuzawa Y. Gastrin activates nuclear factor kappaB (NFkappaB) through a protein kinase C dependent pathway involving NFkappaB inducing kinase, inhibitor kappaB (IkappaB) kinase, and tumour necrosis factor receptor associated factor 6 (TRAF6) in MKN-28 cells transfected with gastrin receptor. *Gut.* 2003 Jun;52(6):813-9.

[17]   Rengifo-Cam W, Umar S, Sarkar S, Singh P. Antiapoptotic effects of progastrin on pancreatic cancer cells are mediated by sustained activation of nuclear factor-{kappa}B. *Cancer Res.* 2007 Aug 1;67(15):7266-74.

[18]   Hiraoka S, Miyazaki Y, Kitamura S, Toyota M, Kiyohara T, Shinomura Y, Mukaida N, Matsuzawa Y. Gastrin induces CXC chemokine expression in gastric epithelial cells through activation of NF-kappaB. *Am J Physiol Gastrointest Liver Physiol.* 2001 Sep;281(3):G735-42.

[19]   Verstak B, Nagpal K, Bottomley SP, Golenbock DT, Hertzog PJ, Mansell A. MyD88 adapter-like (Mal)/TIRAP interaction with TRAF6 is critical for TLR2- and TLR4-mediated NF-kappaB proinflammatory responses. *J Biol Chem.* 2009 Sep 4;284(36):24192-203.

[20]   Scharschmidt E, Wegener E, Heissmeyer V, Rao A, Krappmann D. Degradation of Bcl10 induced by T-cell activation negatively regulates NF-kappa B signaling. *Mol Cell Biol.* 2004 May;24(9):3860-73.

[21]   Liu GY, Liu KH, Zhang Y, Wang YZ, Wu XH, Lu YZ, Pan C, Yin P, Liao HF, Su JQ, Ge Q, Luo Q, Xiong B. Alterations of tumor-related genes do not exactly match the histopathological grade in gastric adenocarcinomas. *World J Gastroenterol.* 2010 Mar 7;16(9):1129-37.

[22]   Kubo-Murai M, Hazeki K, Sukenobu N, Yoshikawa K, Nigorikawa K, Inoue K, Yamamoto T, Matsumoto M, Seya T, Inoue N, Hazeki O. Protein kinase Cdelta binds TIRAP/Mal to participate in TLR signaling. *Mol Immunol.* 2007 Mar;44(9):2257-64.

[23]   Singh P, Wu H, Clark C, Owlia A. Annexin II binds progastrin and gastrin-like peptides, and mediates growth factor effects of autocrine and exogenous gastrins on colon cancer and intestinal epithelial cells. *Oncogene* 2007; 26: 425–4.

[24]   Ewing RM, *et al.,* Large-scale mapping of human protein-protein interactions by mass spectrometry. *Mol Syst Biol.* 2007;3:89.

[25]   Anwar KN, Fazal F, Malik AB, Rahman A RhoA/Rho-associated kinase pathway selectively regulates thrombin-induced intercellular adhesion molecule-1 expression in endothelial cells *via* activation of I kappa B kinase beta and phosphorylation of RelA/p65. *J Immunol.* 2004 Dec 1;173(11):6965-72.

[26]   Scharschmidt E, Wegener E, Heissmeyer V, Rao A, Krappmann D. Degradation of Bcl10 induced by T-cell activation negatively regulates NF-kappa B signaling. *Mol Cell Biol.* 2004 May;24(9):3860-73.

[27]   Gong XG, Lv YF, Li XQ, Xu FG, Ma QY. Gemcitabine resistance induced by interaction between alternatively spliced segment of tenascin-C and annexin A2 in pancreatic cancer cells. *Biol Pharm Bull.* 2010;33(8):1261-7.

[28]   Kabir S, Daar GA. Serum levels of interleukin-1, interleukin-6 and tumour necrosis factor-alpha in patients with gastric carcinoma. *Cancer Lett.* 1995 Aug 16;95(1-2):207-12.

[29]   Forones NM, Mandowsky SV, Lourenço LG. Serum levels of interleukin-2 and tumor necrosis factor-alpha correlate to tumor progression in gastric cancer. *Hepatogastroenterology.* 2001 Jul-Aug;48(40):1199-201.

[30]   Kim DK, Oh SY, Kwon HC, Lee S, Kwon KA, Kim BG, Kim SG, Kim SH, Jang JS, Kim MC, Kim KH, Han JY, Kim HJ. Clinical significances of preoperative serum interleukin-6 and C-reactive protein level in operable gastric cancer. *BMC Cancer.* 2009 May 20;9:155.

[31]   Zabaleta J, Lin HY, Sierra RA, Hall MC, Clark PE, Sartor OA, Hu JJ, Ochoa AC. Interactions of cytokine gene polymorphisms in prostate cancer risk. *Carcinogenesis.* 2008 Mar;29(3):573-8.

[32]   Ikeguchi M, Hatada T, Yamamoto M, Miyake T, Matsunaga T, Fukumoto Y, Yamada Y, Fukuda K, Saito H, Tatebe S. Serum interleukin-6 and -10 levels in patients with gastric cancer. *Gastric Cancer.* 2009;12(2):95-100.

[33]   Liu Y, Cao DJ, Sainz IM, Guo YL, Colman RW. The inhibitory effect of HKa in endothelial cell tube formation is mediated by disrupting the uPA-uPAR complex and inhibiting its signaling and internalization. *Am J Physiol Cell Physiol.* 2008 Jul;295(1):C257-67.

[34]   Oue N, Hamai Y, Mitani Y, Matsumura S, Oshimo Y, Aung PP, Kuraoka K, Nakayama H, Yasui W. Gene expression profile of gastric carcinoma: identification of genes and tags potentially involved in invasion, metastasis, and carcinogenesis by serial analysis of gene expression. *Cancer Res.* 2004 Apr 1;64(7):2397-405.

SUPPLEMENTARY MATERIAL

Supplementary Table 1: A list of 132 significant expression regulators with a SNEA p-value below 0.05 id to synergy between gastrin over-expression and *Helicobacter* infection. All regulators in this table were use analysis to identify what regulators are upstream of others in expression regulatory network. Regulators columns were found to significant in the second SNEA analysis, which used first SNEA regulators as inpu components of NF-κB complex: Functional Class NF-κB had a p-value of 1.40665E-17 in the second SNEA table; IKBKE had a p-value of 9.46645e-11 and regulated 7 proteins in this table. RELB had a p-value of 1.0 this table.

Name	Type	Description	Connectivity in the database	Cell Localization	# of Measured Neighbors	Median change	SNEA p-value i microarray experiment
actin filament	Complex		1737	Cytoplasm	35	2.2	0.0431
adenosine nucleotide receptor	Functional Class		646	Plasma membrane	32	1.8	0.0049
ADIPOQ	Protein	adiponectin, C1Q and collagen domain containing	1203	Extracellular	88	1.8	0.0495
ADORA2B	Protein	adenosine A2b receptor	199	Plasma membrane	17	4.1	0.0155
ALOX15	Protein	arachidonate 15-lipoxygenase	233	Cytoplasm, Plasma membrane	15	1.7	0.0248
ALOX5	Protein	arachidonate 5-lipoxygenase	832	Cytoplasm, Nucleus matrix, Plasma membrane	25	1.8	0.0072
alpha2beta1 integrin	Complex		329	Plasma membrane	10	2.9	0.0184
alphaVbeta6	Complex		120	Plasma membrane	5	2.9	0.0081
APLN	Protein	apelin	253	Extracellular	15	2.1	0.0419
AZU1	Protein	azurocidin 1	132	Cytoplasmic granule, Extracellular	9	3.3	0.0104
B2M	Protein	beta-2-microglobulin	513	Secreted, Plasma membrane	25	2.1	0.0162
CAMKK1	Protein	calcium/calmodulin-dependent protein kinase kinase 1, alpha	35	Cytoplasm, Nucleus	5	3.5	0.0146
CANT1	Protein	calcium activated nucleotidase 1	143	Endoplasmic reticulum membrane, Golgi stack membrane	9	3.5	0.0195
carcinoembryonic antigen	Functional Class		300	Plasma membrane	15	2.2	0.0232

CAT	Protein	catalase	1714	Peroxisome, Cytoplasm	78	2.1	0.0144
CCL14	Protein	chemokine (C-C motif) ligand 14	47	Extracellular	6	3.8	0.0396
CCL5	Protein	T-cell specific protein p288	1077	Extracellular	34	2.2	0.0090
CCND3	Protein	cyclin D3	362	Cytoplasm, Nucleus	7	2.4	0.0457
CD38	Protein	CD38 molecule	432	Plasma membrane	14	2.7	0.0484
CD4	Protein	CD4 molecule	2580	Plasma membrane, Extracellular	165	1.7	0.0315
CD82	Protein	CD82 antigen (R2 leukocyte antigen, antigen detected by monoclonal	241	Plasma membrane, Extracellular	13	2.1	0.0404
CDC37	Protein	cell division cycle 37 homolog (S. cerevisiae)	131	Cytoplasm	7	3.1	0.0114
CEBPE	Protein	CCAAT/enhancer binding protein (C/EBP), epsilon	92	Nucleus	22	2.5	0.0288
CIITA	Protein	class II, major histocompatibility complex, transactivator	322	Nucleus	29	2.2	0.0312
CLU	Protein	clusterin	415	Secreted	15	2.2	0.0361
collagenase	Functional Class		756	Extracellular	7	2.9	0.0062
Csd/Coldshock	Functional Class		25	Nucleus	9	2.1	0.0283
CSF1	Protein	colony stimulating factor 1 (macrophage)	972	Plasma membrane, Extracellular	104	1.9	0.0120
CSN2	Protein	casein beta	242	Secreted	5	2.4	0.0358
CTLA4	Protein	celiac disease 3	615	Plasma membrane	37	1.8	0.0431
CXCL1	Protein	chemokine (C-X-C motif) ligand 1 (melanoma growth stimulating activity, alpha)	552	Extracellular	14	2.3	0.0462
CXCL13	Protein	chemokine (C-X-C motif) ligand 13 (B-cell chemoattractant)	167	Extracellular	10	3.0	0.0457
CXCR4	Protein	neuropeptide Y-Y3	1076	Plasma membrane	24	2.2	0.0307
DDR1	Protein	PTK3A protein tyrosine kinase 3A	190	Plasma membrane, Secreted	22	2.7	0.0030
DDR2	Protein	discoidin domain receptor tyrosine kinase 2	64	Plasma membrane	6	3.6	0.0135
dihydrosphingosine kinase	Functional Class		221		6	3.6	0.0160
DNA topoisomerase II	Functional Class		500	Nucleus	16	2.4	0.0466
EFNB2	Protein	ephrin-B2	125	Plasma membrane,	6	7.7	0.0055

				Extracellular			
ELN	Protein	elastin (supravalvular aortic stenosis, Williams-Beuren syndrome)	571	Extracellular matrix	11	2.0	0.0428
EPHB4	Protein	EPH receptor B4	148	Plasma membrane	14	3.0	0.0421
EPX	Protein	eosinophil peroxidase	137	Cytoplasmic granule	10	4.5	0.0429
ETV1	Protein	hypothetical protein LOC221810	48	Nucleus	6	7.7	0.0276
Fc receptor	Complex		455	Plasma membrane	11	2.2	0.0149
FCER2	Protein	Fc fragment of IgE, low affinity II, receptor for (CD23)	305	Plasma membrane, Secreted	9	4.8	0.0214
FLT3LG	Protein	fms-related tyrosine kinase 3 ligand	240	Plasma membrane, Extracellular	27	2.2	0.0057
Glutathione:hydrogen-peroxide oxidoreductase	Functional Class		854		15	2.5	0.0311
GRM2	Protein	glutamate receptor, metabotropic 2	139	Plasma membrane	5	3.5	0.0161
HABP2	Protein	hyaluronan binding protein 2	86	Extracellular	7	6.1	0.0032
histone H3	Functional Class		709	Nucleus	42	1.8	0.0295
HMGB1	Protein	high-mobility group box 1	701	Nucleus	51	1.7	0.0353
HSP90B1	Protein	heat shock protein 90kDa beta (Grp94), member 1	317	Endoplasmic reticulum lumen, Melanosome, Cytoplasm	8	2.7	0.0371
HSPD1	Protein	spastic paraplegia 13 (autosomal dominant)	645	Mitochondrion matrix, Cytoplasm	37	2.0	0.0252
IFNAR ligand	Functional Class		1753	Extracellular	231	1.7	0.0287
IFNB1	Protein	interferon, beta 1, fibroblast	1012	Extracellular	122	2.2	0.0003
IFNG	Protein	interferon, gamma	4469	Extracellular	574	1.7	0.0041
IgG	Functional Class		2593	Extracellular	75	2.0	0.0069
IL1 family	Functional Class	IL1/18	2567	Extracellular	282	1.7	0.0036
IL10RA	Protein	interleukin 10 receptor, alpha	160	Plasma membrane	18	2.0	0.0383
IL11	Protein	interleukin 11	490	Extracellular	52	1.9	0.0314
IL17A	Protein	interleukin 17A	753	Extracellular	75	2.0	0.0057
IL18	Protein	interleukin 18 (interferon-gamma-inducing factor)	914	Extracellular	75	1.7	0.0457
IL22	Protein	IL-10-related T-cell-derived	215	Extracellular	32	2.2	0.0021

		inducible factor					
IL23	Complex		338	Extracellular	22	3.0	0.0076
IL29	Protein	interleukin 29 (interferon, lambda 1)	70	Secreted	10	3.3	0.0072
IL4R	Protein	interleukin 4 receptor	345	Plasma membrane, Secreted	14	4.5	0.0233
IL8	Protein	interleukin 8	2883	Extracellular	69	1.8	0.0337
IL9	Protein	interleukin 9	229	Extracellular	30	1.7	0.0228
immunoglobulin	Complex		1928	Extracellular	48	2.2	0.0006
inflammatory cytokine	Functional Class		3439	Extracellular	250	1.7	0.0039
interferon	Functional Class		1783	Extracellular	168	1.8	0.0014
ISGF3	Complex		62	Nucleus	15	2.8	0.0229
ITGAM	Protein	integrin, alpha M (complement component 3 receptor 3 subunit)	858	Plasma membrane	18	3.0	0.0085
ITGAX	Protein	integrin, alpha X (complement component 3 receptor 4 subunit)	238	Plasma membrane	8	1.9	0.0378
ITGB2	Protein	integrin, beta 2 (complement component 3 receptor 3 and 4 subunit)	952	Plasma membrane	20	4.6	0.0022
JNK	Functional Class	JUN kinase	659	Cytoplasm	23	1.8	0.0453
LBP	Protein	lipopolysaccharide binding protein	187	Secreted	10	5.0	0.0271
LIF	Protein	leukemia inhibitory factor (cholinergic differentiation factor)	875	Extracellular	123	1.7	0.0416
LIFR	Protein	leukemia inhibitory factor receptor alpha	137	Plasma membrane, Secreted	10	2.6	0.0195
Lipoprotein(a)	Complex		437	Extracellular	19	2.2	0.0245
LTBR	Protein	lymphotoxin beta receptor (TNFR superfamily, member 3)	183	Plasma membrane	21	2.0	0.0308
LYN	Protein	v-yes-1 Yamaguchi sarcoma viral related oncogene homolog	610	Plasma membrane, Cytoplasm, perinuclear region, Golgi apparatus	16	2.9	0.0177
MAP2K4	Protein	mitogen-activated protein kinase kinase 4	450	Cytoplasm	33	1.8	0.0469
MAP2K7	Protein	mitogen-activated protein kinase kinase 7	198	Nucleus, Cytoplasm	21	2.1	0.0148

MAP3K14	Protein	mitogen-activated protein kinase kinase kinase 14	162	Cytoplasm	7	2.2	0.0407
MBL2	Protein	mannose-binding lectin (protein C) 2, soluble (opsonic defect)	432	Secreted	12	2.7	0.0424
membrane attack	Complex		223	Extracellular	12	5.0	0.0016
MKK3/6	Functional Class	mitogen-activated protein kinase kinase 3/6	114	Cytoplasm	7	4.8	0.0068
nAChR	Functional Class	neuronal nicotinic acetylcholine receptor	888	Plasma membrane	18	2.1	0.0349
NFKB1	Protein	nuclear factor of kappa light polypeptide gene enhancer in B-cells 1 (p105)	518	Nucleus, Cytoplasm	62	1.8	0.0458
NFKBI	Functional Class	NF kappa B inhibitor	659	Cytoplasm	77	1.9	0.0223
NFKBIA	Protein	nuclear factor of kappa light chain gene enhancer in B-cells inhibitor, alpha	1034	Cytoplasm, Nucleus	46	2.1	0.0087
NIK	Functional Class	mitogen-activated protein kinase kinase kinase 14	273	Cytoplasm	27	2.2	0.0236
NTRK3	Protein	neurotrophic tyrosine kinase, receptor, type 3	184	Plasma membrane	6	3.0	0.0439
OAS1	Protein	2',5'-oligoadenylate synthetase 1, 40/46kDa	125	Mitochondrion, Nucleus, Microsome, Endoplasmic reticulum, Cytoplasm	5	4.8	0.0165
OSM	Protein	oncostatin M	531	Extracellular	112	1.8	0.0087
P2RY11	Protein	purinergic receptor P2Y, G-protein coupled, 11	60	Plasma membrane	8	3.0	0.0243
PDE4A	Protein	phosphodiesterase 4A, cAMP specific	430	Cytoplasm, perinuclear region, Cell projection, ruffle membrane	12	3.0	0.0275
PEA15	Protein	phosphoprotein enriched in astrocytes 15	126	Cytoplasm	7	3.5	0.0497
phospholipase A2	Functional Class	phospholipase A2	1924	Cytoplasm	38	1.8	0.0359
PLA2G4A	Protein	phospholipase A2, group IVA (cytosolic, calcium-dependent)	293	Cytoplasm, Cytoplasmic vesicle	10	3.1	0.0460
PMNL proteinase	Functional Class		183		11	4.5	0.0112

POU2AF1	Protein	POU class 2 associating factor 1	88	Nucleus	13	1.8	0.0460
PPP2CB	Protein	protein phosphatase 2 (formerly 2A), catalytic subunit, beta isoform	157	Cytoplasm, Nucleus, Centromere, cytoskeleton, spindle pole	5	3.7	0.0116
PTGER1	Protein	prostaglandin E receptor 1 (subtype EP1), 42kDa	226	Plasma membrane	17	2.2	0.0172
PTGER2	Protein	prostaglandin E receptor 2 (subtype EP2), 53kDa	391	Plasma membrane	56	2.3	0.0069
PTMA	Protein	prothymosin, alpha	221	Nucleus	12	2.9	0.0478
PYCARD	Protein	PYD and CARD domain containing	174	Cytoplasm	7	3.1	0.0201
pyrogen	Functional Class		147	Extracellular	10	4.5	0.0083
PYY	Protein	peptide YY	316	Extracellular	8	2.9	0.0417
RAG2	Protein	recombination activating gene 2	123	Nucleus	6	3.8	0.0368
REL	Protein	v-rel reticuloendotheliosis viral oncogene homolog (avian)	629	Nucleus	83	1.8	0.0109
S100A12	Protein	S100 calcium binding protein A12	97		6	4.8	0.0064
SAP30	Protein	Sin3A-associated protein, 30kDa	181	Nucleus	23	2.1	0.0362
SELP	Protein	selectin P (granule membrane protein 140kDa, antigen CD62)	957	Plasma membrane	11	2.2	0.0144
serine protease inhibitor	Functional Class		340		6	6.5	0.0229
SERPINA4	Protein	serpin peptidase inhibitor, clade A (alpha-1 antiproteinase, antitrypsin), member 4	126	Secreted	9	4.5	0.0258
SERPING1	Protein	serpin peptidase inhibitor, clade G (C1 inhibitor), member 1	261	Extracellular	5	4.8	0.0137
SNCA	Protein	synuclein, alpha (non A4 component of amyloid precursor)	632	Cytoplasm, Plasma membrane, Nucleus	17	2.0	0.0421
SOCS1	Protein	suppressor of cytokine signaling 1	562	Cytoplasm	42	2.0	0.0338
SOCS3	Protein	suppressor of cytokine signaling 3	682	Cytoplasm	33	1.8	0.0277
SOX8	Protein	SRY (sex determining region Y)-box 8	36	Nucleus	5	4.9	0.0464
SPARC	Protein	secreted protein, acidic, cysteine-rich (osteonectin)	455	Extracellular matrix, basement membrane	32	2.2	0.0304
SPN	Protein	sialophorin	229	Plasma membrane	11	2.2	0.0291
superoxide dismutase	Functional Class		1753		131	1.8	0.0076
TLR2	Protein	toll-like receptor 2	981	Phagosome, Plasma	88	1.9	0.0006

				membrane			
TMSB4X	Protein	thymosin, beta 4, X-linked	192	Cytoplasm, cytoskeleton	13	2.1	0.0224
TNFRSF13C	Protein	tumor necrosis factor receptor superfamily, member 13C	93	Plasma membrane	9	2.1	0.0342
TNFRSF9	Protein	tumor necrosis factor receptor superfamily, member 9	247	Plasma membrane	23	2.2	0.0142
TNFSF18	Protein	tumor necrosis factor (ligand) superfamily, member 18	73	Plasma membrane, Extracellular	12	2.1	0.0476
xanthine oxidase	Functional Class	xanthine oxidase	188		16	4.5	0.0038
XDH	Protein	Xanthine dehydrogenase/oxidase	885	Peroxisome, Cytoplasm, Secreted	27	2.0	0.0177
ZNF384	Protein	zinc finger protein 384	28	Nucleus	5	2.7	0.0354

CHAPTER 6

Sub-Network Enrichment and Cluster Analysis Reveal Possible Pathways for Cetuximab Sensitivity

Mikhail A. Pyatnitskiy[*], Maria A. Shkrob, Nikolai D. Daraselia and Ekaterina A. Kotelnikova

Ariadne Genomics Inc., Rockville, MD, USA

Abstract: Patient stratification or, a personalized approach to medical treatment, is a promising approach in modern medicine. Finding biological patterns within a group of patients with the same diagnosis could lead to more precise and effective therapies. To address this issue it is necessary to reveal different mechanisms within the same disease, to find new biomarkers, and to develop new diagnostic tests that would distinguish patients from different subgroups.

Cancer sub-typing based on clustering of individual patient gene expression profiles has been widely used for various types of cancer. Here we propose a new approach which includes the consecutive use of Sub-network enrichment analysis algorithm (SNEA) for individual differential expression profiles and biclustering of found expression regulators and samples.

We analyzed nine publicly available microarray datasets with data from patients suffering from colorectal cancer as compared to healthy donors, including one dataset containing supplementary information on patient response to anti-EGFR therapy with cetuximab. We have identified several patient subtypes characterized by specific regulatory clusters (pathways) and mapped the data about cetuximab response onto the heat map of pathway activity for each patient. We found that the most prominent mechanism that distinguished responders from non-responders is dependent on regulators from the TGF-β/SMAD pathway and corresponds to the epithelial-to-mesenchymal transition (EMT).

Keywords: Bioinformatics, colorectal cancer, microarray, gene expression, subnetwork enrichment analysis, patient stratification, cluster analysis, cetuximab, EGFR, KRAS, regulator, epithelial-to-mesenchymal transition, pathway analysis, biomarker, text-mining, pathway studio, MedScan, CRC, differential profile, biclustering.

*Address correspondence to Mikhail A. Pyatnitskiy:** Ariadne Genomics, Rockville, USA, and Institute of Biomedical Chemistry, RAMS, Moscow, Russia; E-mail: mpyat@ariadne.net

Anton Yuryev and Nikolai Daraselia (Eds)

INTRODUCTION

Colorectal cancer (CRC) is among the most prevalent cancers in both men and women. Each year roughly one million of new CRC cases are diagnosed worldwide [1]. Approximately half of all patients with CRC will develop metastases and the majority of these patients will die from the cancer [2]. About 141,000 new cases and 50,000 deaths from CRC are expected in 2011 in the U.S. [3]. Although CRC mortality has been progressively declining since 1990 at a rate of about 3% a year, it remains the second most common cause of cancer death in the U.S. [4]. The approval of several new drugs in the U.S. such as oxaliplatin (Eloxatin, Sanofi-Aventis, 2002), bevacizumab (Avastin®, Genentech/Roche, 2004) and especially anti-EGFR therapy (cetuximab and panitumumab) has increased the average median survival rate for metastatic CRC (stages III and IV) by more than two years.

Cetuximab (Erbitux®, ImClone Systems, 2004) and panitumumab (Vectibix®, Amgen, 2006) are monoclonal antibodies that bind with EGFR to prevent ligand binding and activation of downstream signaling pathways activated during cancer cell proliferation, invasion, metastasis, and stimulation of neovascularization. These drugs are approved by the FDA for the CRC treatment in the refractory disease setting.

Colorectal cancer is a major problem in industrial countries. Together with lung, breast and prostate cancers they constitute the majority of lethal cancers. However, CRC is much less widespread in the developing world [5]. These differences suggest that life style and nutrition play important roles the in etiology of a disease. Hence, CRC can be viewed as a perspective disease for the development of personalized medicine, in which individualized medical treatment and care are prescribed based on personal and genetic variation using advances in technology such as a biomarker test.

One example of a biomarker test already used in the U.S. and Europe to predict therapeutic response of anti-cancer agents is the KRAS mutation test for Erbitux® and Vectibix®. Currently, KRAS mutation testing is recommended by the National Comprehensive Cancer Network before starting EGFR-targeted therapy in both metastatic CRC and advanced non-small cell lung cancer patients [6,7]. This real-time PCR DNA test can identify the status of the KRAS gene: wild-type (WT) or mutant form. Based on CRYSTAL trial results, about 60% of metastatic CRC patients have WT KRAS tumors and therefore are eligible for anti-EGFR therapy [6,9]. However, approximately 55-60% of patients with wild-type KRAS do not benefit from anti-EGFR therapy, which is also very toxic and highly expensive [9]. For example, the most common side effect of anti-EGFR therapy is skin toxicity [10].

The primary efforts of medical research focus on improving chances for the successful outcome of a therapy. It can be achieved through stratification of patients using biomarker profiling prior to drug treatment. Understanding the molecular mechanisms of disease in every patient moves us closer to finding the right drug for the right patient.

Sub-Network Enrichment Analysis: The most common approach for selecting candidate biomarkers is looking among genes differentially expressed between the condition for diagnosis and a corresponding "control" condition, *e.g.*, between disease and normal tissues, or between drug responders and non-responders. If no single gene can distinguish the conditions, machine-learning or clustering algorithms can be used to find complex multi-gene signature(s) with improved sensitivity and specificity [11,12] The potential drawback of this approach is a possibly small predictive power: *i.e.*, a gene signature working for a certain group of patients may perform poorly for the next cohort. This can be due to different molecular mechanisms of disease or due to poor scalability ability of the classifier.

The traditional approach analyzes an expression dataset at the level of individual genes, and it is common for most genes from a pathway P to have poor p-values. As a consequence, the most differentially expressed genes are unlikely to be enriched by members of P, and thus P cannot be considered as an affected pathway. However, a correlated deviation of 20 genes from one pathway by 10% is a much more statistically significant event than the differential expression of a single gene even by 200%. Therefore, it is desirable to develop a method which will identify events when genes related to the same pathway exhibit consistent changes in the expression pattern, even if their individual changes are small.

Here we describe an enhancement to the traditional approach that involves the additional step of finding statistically significant regulators of differential expression. Instead of looking only at differentially expressed genes, we first identify transcriptional regulators and select biomarkers downstream from them. We used Sub-network enrichment analysis(SNEA) [13] to identify major expression regulators. SNEA is a variation of the more commonly used gene set enrichment analysis (GSEA) algorithm.

Unlike GSEA, which uses predefined collection of gene sets, SNEA uses a global protein-gene expression regulatory network extracted from scientific literature to generate a collection of gene sets, each representing all known downstream expression targets of one protein (SNEA "seed"). Thus, if downstream expression targets of the "seed" protein are enriched with differentially expressed genes, then the "seed" protein can be interpreted as a key regulator of an observed differential expression profile. SNEA implemented in Pathway Studio calculates p-values indicating enrichment of each sub-network with differentially expressed genes using the non-parametric Mann-Whitney test.

SNEA uses a global expression regulatory network extracted from the entire PubMed and from full-text articles in more than 60 open-access full-text journals, using MedScan information extraction technology [14,15]. Currently, the network contains more than 165,000 unique protein-gene expression regulation relations. It represents up-to-date scientific knowledge without bias and always includes the data from the latest research papers.

In other words, we use the MedScan extracted regulatory network to find key transcription regulators of the differential expression profile. The regulators provide unique insights into the mechanisms and biological pathways that are responsible for clinical outcome. The analysis of regulators allows for the rational selection of biomarkers that are specifically downstream from the identified seeds and may also increase generalization ability of the entire test.

Approach. The sub-typing of cancer based on the clustering of gene expression profiles is still being researched [16-19]. One of the pioneering publications with an application to colon cancer was the study by Alon *et al.,* [20]. We have extended the traditional approach by calculating a differential expression profile compared to normal tissue and by applying the SNEA algorithm to transform the large number of less informative differential profiles into a much smaller set of expression regulators that presumably lead directly to the individual mechanisms of cancer in a patient.

To identify potential biomarkers of the anti-EGFR response we have used the following strategy:

- We have aggregated all publicly available gene expression datasets for colorectal cancer biopsies and normal colon tissues in order to calculate a cancer expression profile for each CRC sample. We applied the SNEA algorithm to identify key significant regulators that affect gene expression separately in every patient. We then clustered significant regulators to identify molecular sub-types of colorectal cancer.

- We have superimposed our cancer profiles for patients with known cetuximab sensitivity information onto the global map of colorectal cancer sub-types in order to find a correlation between cancer sub-types and anti-EGFR treatment sensitivity.

- We have compared public gene expression profiles of CRC patients who responded to cetuximab treatment to patients with no response. We have applied SNEA to look at key significant regulators of genes differentially expressed between responders and non-responders and performed an extensive literature analysis to build a preliminary pathway of anti-EGFR treatment sensitivity.

RESULTS

The entire colon cancer dataset was divided into two parts: training and test sets in order to prevent over-fitting. The training dataset included CRC samples from all studies, except for GSE5851, which was used as a test dataset. Key regulators were identified by SNEA on the training set and intersected with regulators obtained on the test set. Then we clustered samples from the test set using previously identified regulators as features. We found group of patients not responding to cetuximab therapy, which is characterized by activation of TGF-β-SMAD pathway.

Our training dataset contained 703 samples of human CRC. For every sample normalized on normal colon tissue expression we ran the SNEA algorithm in Pathway Studio with the following parameters: type of seed was "protein", "functional class" or "complex", seed neighbor type was "Protein" and types of relation links were either "Expression" or "Promoter Binding." We found 695 regulators, which were significant in at least one sample. To eliminate SNEA regulators that were responsible for natural variability in gene expression between individual patients we defined them as "rare" regulators, which were encountered in fewer than 20 samples total. We found 522 regulators to fit this definition and excluded them from further analysis. We also removed four patient samples which contained only rare regulators. Finally, our training dataset included 173 regulators and 699 samples.

Expression Regulators Clustering: An analysis of significant expression regulators rather than of differentially expressed genes greatly facilitates biological interpretation of gene expression microarray data. While SNEA allows quick identification of the regulators, a higher level of understanding of the mechanisms' underlying disease is still desirable. A relevant and adequate description of cell functioning can be achieved at the level of pathways to explain differences between normal and disease conditions. The list of hundreds SNEA regulators is difficult to interpret since many regulators have many target genes in common and cooperate together to regulate gene expression. We believe that considering groups of similar regulators should simplify the further interpretation of SNEA results. Finding groups of similar regulators can serve as the first step in the bottom-up automatic reconstruction of pathways.

The similarity between two regulators can be defined as a percentage of common targets, which can be calculated as Jaccard distance between two sets of targets. We calculated the Jaccard distance similarity matrix of 173 significant regulators identified as "non-rare" in the training dataset and performed their clustering using Ward's hierarchical clustering method.

To obtain a reasonable number of regulator clusters we used the L-shaped method [21]. This approach tries to find a knee in a "number of clusters *vs.* clustering cut-off" curve. The curve is approximated with two piecewise linear regressions to achieve the smallest possible total sum-of-square error. The intersection of two linear approximations gives an estimate of reasonable number of clusters. The L-shaped method suggested the optimal clustering cutoff of 1.6. Cutting the clustering dendrogram at this level produced 14 clusters of regulators, which are shown in Fig. **1**.

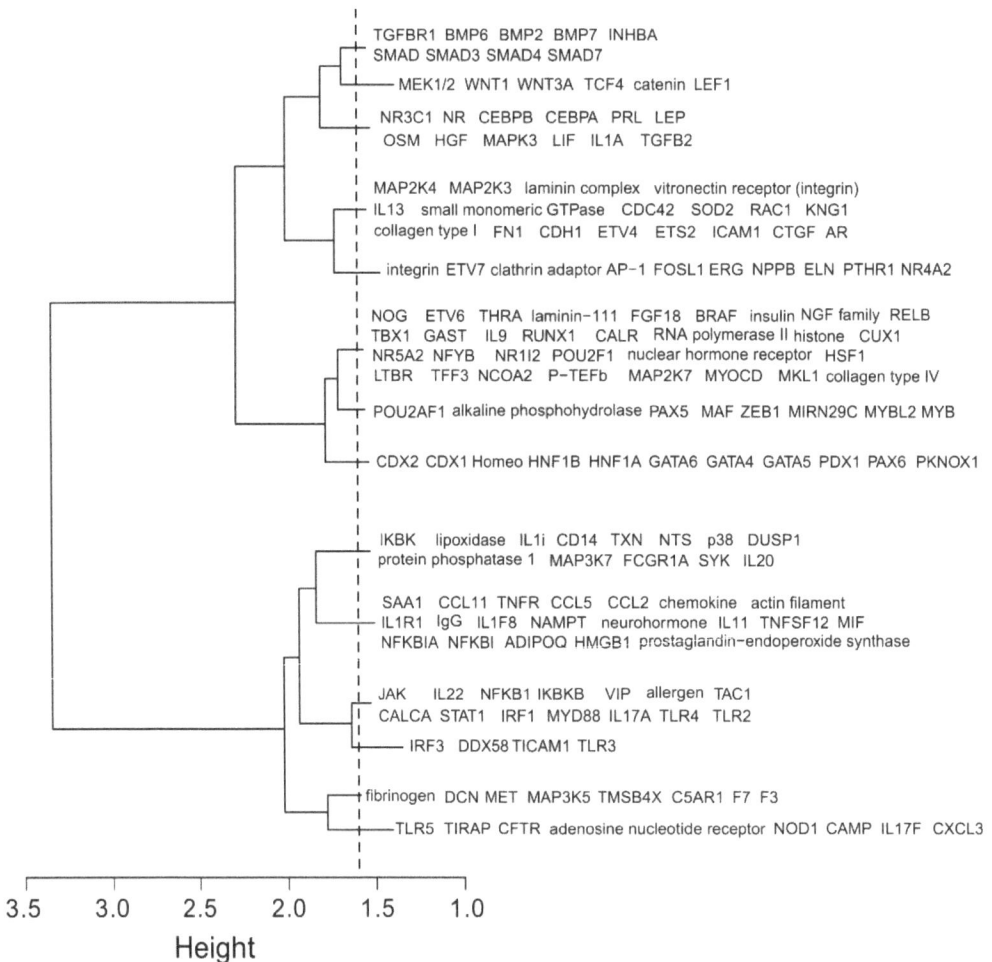

Figure 1: A hierarchical clustering of 173 significant regulators identified using SNEA on 699 CRC samples. Similarity between each pair of regulators is estimated as percentage of their common targets. The L-shaped method revealed an optimal dendrogram cutting equal to 1.6, which gives 14 clusters of regulators.

The resulting clusters of regulators can be interpreted as individual pathways. For example, the cluster at the top of the dendrogram in Fig. **1** represents the TGF-β/SMAD signaling pathway, including receptor (TGFBR1), ligands (BMP2, BMP6, BMP7), transcription factors activating (SMAD3) or inhibiting (SMAD7) expression of target genes. INHBA is involved in activin production and also is a member of the TGF-β superfamily.

Several studies suggest that there is a link between blood coagulation and cancer progression [22,23]. Tumor cells activate blood coagulation by expressing procoagulant substances and fibrinolytic factors. A correlation between global fibrinolytic capacity and the presence of metastases was shown for colorectal cancer [24]. We found further support for this correlation in a cluster of regulators containing fibrinogen, coagulation factors III and VII. Decorin, another member of this cluster, is also related to fibrin clotting [24].

We propose that identified clusters of regulators can be used to discover molecular mechanisms responsible for a specific condition and to allow for the rational selection of individualized therapeutic treatment. We demonstrate this in the next section using an example of CRC patients treated with cetuximab and showing different types of responses to this therapy.

Sample Clustering: We have applied the same type of SNEA analysis to our test set (GSE5851) to see if cetuximab responders belong to any specific sub-type of colorectal cancer. The differential gene expression profiles of CRC patients *vs.* normal colon tissue have been calculated and SNEA identified 383 regulators which were significant in at least one sample. Interestingly, we have found few liver-specific regulators in metastatic CRC patient profiles, either due to tissue contamination of metastatic biopsies or due to re-differentiation of metastatic tumors inside liver – the site of biopsies used in GSE5851.

For consistency, we have ignored all regulators not identified in the training set (699 purely colorectal differential profiles), which resulted in a list of 118 unique significant regulators. They have been assigned to 14 clusters identified previously in the training set. The detailed composition of 14 clusters of SNEA regulators along with their functional description is given in Table **1**.

Table 1: Clusters of significant SNEA regulators identified in both training and test sets.

Cluster	Significant Regulators	Description
C1	CDX2,Homeo,GATA6,HNF1A,HNF1B,CDX1,GATA4	Liver and intestine-specific gene regulation
C2	IL17A,TAC1,JAK,TLR2,IL22,CALCA,IKBKB,TLR4,MYD88	Inflammatory and immune response
C3	laminin-111,GAST,P-TEFb,nuclear hormone receptor, histone, PAX6, TBX1, PDX1, PKNOX1, NFYB, insulin, NR5A2, IL9, POU2AF1,NR1I2,NGF family, MAP2K7,CALR,NCOA2,POU2F1,MYOCD,THRA	Adipocytokine signaling, gonadotrope cell activation
C4	CCL5,IL1R1,chemokine,IL11,CCL2,MIF,HMGB1,IL1F8 prostaglandin-endoperoxide synthase,TNFR	Immunoregulatory and inflammatory processes.
C5	CAMP,adenosine nucleotide receptor,CFTR,FGF18,TLR5,IL17F	Pathogen recognition and activation of immune responses
C6	OSM,TGFB2,IL1A,PRL,HGF,CEBPB,LIF,NR3C1,LEP, MAPK3,NR,CEBPA	Regulation of body weight
C7	WNT3A,TCF4,LEF1,MEK1/2	Wnt-signalling
C8	F7,fibrinogen,F3,DCN,MET,C5AR1	Coagulation
C9	SMAD4,SMAD,INHBA,SMAD3,SMAD7,BMP7,BMP2, TGFBR1	TGF-β/SMAD pathway
C10	IL1i,CD14,lipoxidase,IL20,MAP3K7,SYK,p38,TXN	Cell cycle regulation
C11	laminin complex,collagen type I,CTGF,MAP2K3,MAP2K4,IL13,ETS2,small monomeric GTPase,ETV4,CDH1,FN1,KNG1,ICAM1,vitronectin receptor (integrin)	Connective tissue, epithelial to mesenchymal transition
C12	MIRN29C,ZEB1,MYBL2,Maf,MAF	Regulation of transcription
C13	NR4A2,PTHR1,FOSL1,NPPB	Response to hypoxia
C14	IRF3,DDX58	Regulation of interferon production, response to virus

To cluster cetuximab-treated patients into groups according to their gene expression downstream from the regulators, we define the similarity measure between two patients. Let's denote r_i - vector of log-ratios of all genes, controlled by i-th regulator. For each regulator we computed the median of log-ratios of downstream genes multiplied by number of downstream genes:

$$K_i = median(r_i) \times |r_i|$$

The value K_i reflects the contribution of i-th regulator into a global pattern of differential gene expression. In order to elicit a contribution for entire j-th cluster of regulators we summed the corresponding K-values for all N_j regulators belonging to the cluster:

$$C_j = \sum_{i=1}^{|N_j|} K_i$$

Hence, each sample can be characterized by specifying $C_j, j \in [1; 14]$ values for all clusters of regulators. We performed Ward's hierarchical clustering using Pearson's correlation coefficient as a distance measure between C-vectors of every patient in order to group patients with similar activity profiles of regulator clusters. The resulting heat map of cluster activity in 80 cetuximab-treated CRC patients is shown in Fig. **2**. The patient response to cetuximab therapy is color-coded and shown above the heat map.

The heat map from Fig. **2** shows that 80 cetuximab-treated patients can be grouped into four clusters: 4 partial responders in first left cluster and 2 complete responders in the third cluster. However, three clusters have a heterogeneous proportion of responders and non-responders except for second left cluster shown in Fig. **2** by the outline. This cluster consists mostly of non-responders: 15 out of 17 patients developed progressive disease after the cetuximab treatment and, for 2 patients, the disease had stabilized.

To gain more understanding about the molecular mechanisms underlying grouping of patients from Fig. **2** with respect to progressive disease, we compared the contribution of different features into patient clustering, *i.e.*, clusters of regulators. We found that C9 and C11 regulator clusters are mainly responsible for the clustering of the non-responders' patient group mentioned previously. We computed 95% confidence intervals for mean values of C_9 and C_{11} for samples within and outside the corresponding cluster. Mean $\overline{C_9}$-value for samples within the C9 cluster was 152.9±40.8 and for all other samples (not belonging to cluster C9) mean $\overline{C_9}$-value was 11.9±6.5. Mean $\overline{C_{11}}$-values for samples within and outside the cluster C11 were 132.7±37.5 and 34.4±10.0 correspondingly. In both cases, p-values for the Mann-Whitney U-test were less than 0.001.

Next we turned to a biological interpretation of C9 and C11 clusters of regulators. Cluster C9 contains genes belonging to the TGFβ1/SMAD pathway. The deregulated TGF-β signaling pathway has long been known to be involved in the pathogenesis of human solid tumors and in the development of colorectal cancer in particular [26,27,28]. TGF-β decreases proliferation of epithelial cells by inducing CDK inhibitors and by promoting cell-matrix interactions. Several key proteins involved in tumor metastasis, such as Snail, Slug (zinc finger transcriptional repressors) and SMAD-interacting protein 1 (SiP1), are positively regulated by TGF-β. Any disruption in TGF-β signaling, either by mutational inactivation or by downregulation of expression of any of components involved in TGF-β signaling leads to tumor development. On the contrary, at the advanced cancer stage, TGF-β becomes pro-inflammatory and pro-metastatic [29]. The molecular details of this switch remain to be elucidated. Currently, it is believed, however, that the epithelial to mesenchymal transition (EMT) known to be induced by TGF-β is suspected to play a significant role in this switch.

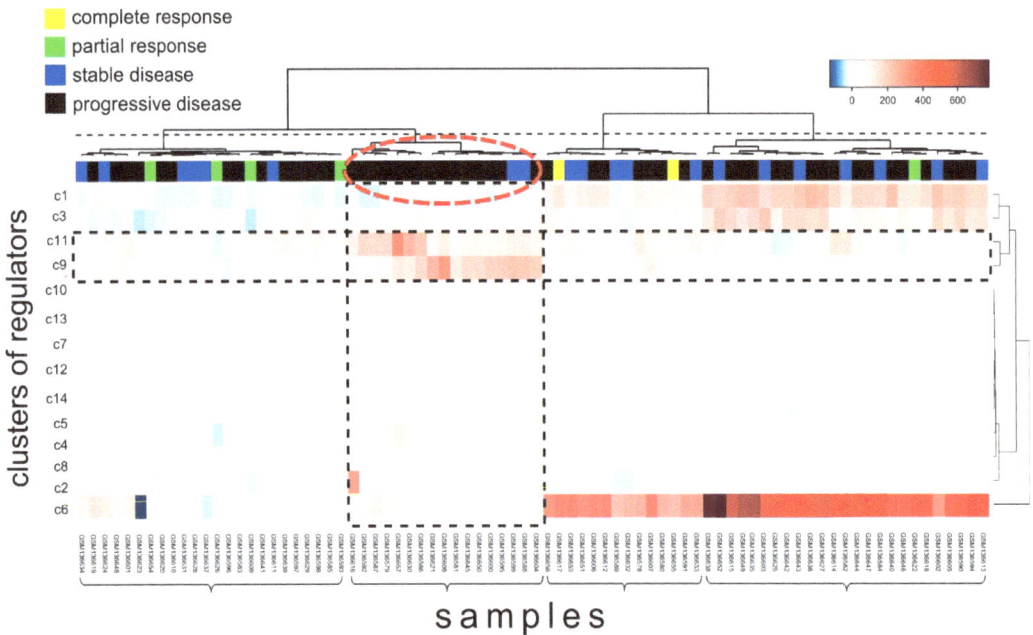

Figure 2: The heat map of activity of 14 clusters in 80 CRC patients treated with cetuximab. The heat map is color-coded by the cetuximab response. The outlined sample cluster consists of mostly of cetuximab non-responders and is characterized by elevated activity of the TGF-β/SMAD pathway (c9) as well as by epithelial to mesenchymal transition markers (c11).

Regulators from the C11 cluster are involved in EMT activation. Notably, cluster C11 contains genes directly involved in EMT: epithelial markers (laminin complex, collagen I, E-cadherin), mesenchymal markers (integrin, fibronectin,

CTGF). Other seeds from the C11 cluster (ETV4, ETS2) activate matrix metalloproteinase genes, associated with the invasion and metastasis of tumor cells. In a recent study [30] it was shown that the expression of MAP2K4 related to tumor invasion resulted from an EMT-like morphological change. Taken together, our findings support the hypothesis that the reason for non-sensitivity to anti-EGFR treatment is the TGF-β–induced epithelial-to-mesenchymal transition.

Analysis of Cetuximab Responders *vs.* Non-Responders: To better understand mechanisms of anti-EGFR treatment sensitivity, we calculated the differential expression profile between patients who showed partial or complete response *vs.* progressive disease patients from GSE5851 and used SNEA to find major expression regulators. Table **2** shows identified significant regulators responsible for differential gene expression between drug responders and non-responders.

Table 2: Significant regulators identified by SNEA of differential profiles for patients showing partial or complete response *vs.* progressive disease (GSE5851).

Significant regulators	Size	Measured Targets	p-value
FOXM1	7	ESR1,CYP3A4,SFTPB,SFTPC,SFTPD,SCGB1A1,NKX2-1	0.000867
FGF10	6	ELF5,LFNG,SFTPB,SFTPC,HOXB13,NKX3-1	0.002451
HIF1A	12	ESR1,FOXP3,AGER,SLC6A4,NT5E,SLC11A1,PDPK1,AQP4,TGFB1,ENO1,AGTR2,AREG	0.003853
CD40LG	5	IL17A,IL15,SERPINB9,SLC6A4,TNFRSF11B	0.007421
Interferon	13	ESR1,IL15,BAX,RUNX2,OPTN,NF1,IL13,MME,DHFR,TGFB1,TNFRSF11B,ROCK1,NEU1	0.011263
NKX2-1	6	SFTPB,SFTPC,AGER,SFTPD,TGFB2,SCGB1A1	0.014028
IL1 family	15	IL11,BAX,PTPRG,SLC6A4,NT5E,CRHR2,STAR,PDGFA,IL17A,CD1A,SFTPB,AQP4,TNFRSF11B,TGFB1,AGTR2	0.016867
TGFB1	42	THRA,IL11,SLC6A2,IL13RA1,ESR1,FOXP3,PAX2,CYP19A1,SFTPC,RARG,CSRP2,MME,STAR,NR2F2,CCNG1,ALPI,COL11A2,TPM1,LAMA1,PLS3,CHRM2,ECM1,EGFR,BAX,GFAP,RUNX2,AGER,AIFM1,PSIP1,LALBA,CALCR,NKX2-1,PDGFA,THBS1,IL17A,CCR4,SFTPB,TNFRSF11B,KCNMA1,TGFB2,AREG,ADRBK1	0.019462
AGT	23	SLC6A2,EGFR,PAX2,NR0B1,BAX,MTUS1,AIFM1,EREG,MC2R,CYP11B1,PDGFA,THBS1,CREM,PDE3A,TRPM6,UBE2I,TPM1,DHFR,TNFRSF11B,TGFB1,ZFP36L1,AGTR2,ROCK1	0.028745
NCOA1	6	ESR1,SFTPB,PTCRA,RNF14,SS18,MC2R	0.030729

NKX2.1/FOXM1 and FGF10 were top scoring regulators in the differential response profile. We also observed the differential expression of TGFB1 targets. NKX2.1 is known to be derepressed by the canonical TGF-B/SMAD pathway [31,32,33] and by activated ERK1/2 kinase [34]. Reciprocally, NKX2.1 was shown to inhibit EMT caused by TGF-β and to restore epithelial phenotype in lung adenocarcinoma cells. This effect was accompanied by downregulation of TGF-β target genes, including presumed regulators of EMT, such as Snail and Slug [35]. Similarly, FOXM1 has also been shown to suppress EMT [36]. Thus, hyperactivity of the FOXM1/NKX2.1 pathway in patients sensitive to cetuximab indicates either downregulation of the TGF-β pathway activity or non-canonical activity of the TGF-β pathway not involving NKX2.1 and FOXM1.

To gain further insights into the mechanisms of resistance to EGFR inhibition, we have analyzed two public datasets: GSE4342 [37] and GSE20854 [38]. These experiments profile a response to EGFR inhibition in cancer cell lines. We analyzed the sensitivity profile, *i.e.*, genes differentially expressed between cancer cell lines resistant to EGFR inhibition and EGFR-sensitive cell lines, using SNEA and inspected significant regulators of differential expression.

Table 3: Significant regulators of differential profiles identified by SNEA that are related to anti-EGFR drug sensitivity (GSE4342 and GSE20854).

Significant Regulators	Size	Measured Targets	p-value
GSE4342 – Gefitinib effect on sensitive/resistant NSCLC cell lines			
HGF	16	CDH1,HMOX1,CLDN3,OCLN,CDH3,LAMC2,BAX,DUSP1,ITGA2,CLDN7,CLDN4,F3,KRT19,SERPINE1,CCND1,VIM	0.006877
INHBA	7	CDH1,CDH2,OCLN,FGFR2,SERPINE1,MYH10,VIM	0.016102
FOXA2	6	TACSTD1,AGR2,AQP3,WNT7B,FOXA1,MAP1B	0.01889
BMP7	6	CDH1,HMOX1,STS,SERPINE1,NR3C2,VIM	0.022266
SNAI1	5	CDH1,CLDN3,OCLN,VDR,SERPINE1	0.029744
GSE20854 – Endometrial cancer cells responsive/sensitive to iressa - treated with EGF and iressa			
ZEB1	13	COL2A1,CDH1,COL1A1,TP73,FN1,MIRN21,HPGD,CDH2,MMP1,TP53,ATP1A1,CCNG2,VDR	0.000243
laminin-111	11	FGF2,IL8,PLAU,CXCR4,CA2,MMP9,MMP1,LAMA1,PODXL,JAG1,FGFR1	0.001488
HOXD3	8	DSP,CDH1,ITGB3,PLAU,JUP,ITGA3,JAG1,TGFBI	0.002053

Table 3: cont….

SOX9	23	COL2A1,CTNNBIP1,NES,BAG1,CEBPB,COL4A2,VPRBP,RUNX2,SOX6,CEBPD,PRKCA,NFATC1,MATN1,LXN,CDH2,MITF,SOX5,TP53,COL9A2,ERBB3,CCND1,PRAME,SOX17	0.002077
MET	11	CASP3,CDH1,CTGF,THBS1,IL8,PTGS2,MMP1,FN1,RELN,TNFSF10,BIRC4	0.002464

Table **3** shows that, in both experiments, significant regulators either belong to the TGF-β pathway (INHBA, BMP7) or are involved in EMT transcriptional regulation (ZEB1, SNAIL1). Interestingly, the most frequently occurring differentially expressed genes downstream of all key regulators in both sensitivity profiles are known EMT biomarkers (CDH1, CDH2, VIM, OCLN, SERPINE1, CLDN3, FN1, *etc.*). The role of EMT in cetuximab sensitivity has been suggested recently from the transcriptomics analysis of epidermoid cancer cells [39]. Our results support this hypothesis and points specifically to the FOXM1/NKX2.1 pathway as an intersection point.

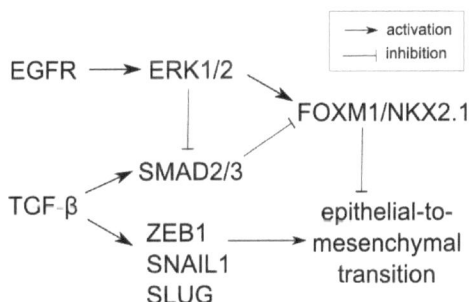

Figure 3: A simplified model for cetuximab resistance based on crosstalk between TGF-β and EGFR pathways. The inhibition of the EGFR pathway leads to derepression of EMT which eventually can lead to tumor redifferentiation and its independence from EGFR activity for growth.

We propose that resistance to EGFR inhibition is related to non-canonical TGFR signaling that is possibly related to EMT. Our model for cetuximab resistance is shown in Fig. **3** is based on the known crosstalk between TGF-β and EGFR pathways which occurs on the level of SMAD2/3 ERK and has NKX2.1 as a focal point. We have performed an extensive literature search and have built a detailed pathway showing the possible mechanism of cetuximab-resistance (Fig. **4**) in which we have highlighted entities that have identified in the C9 cluster (TGF-β/SMAD signaling) and the C11 cluster (epithelial-to-mesenchymal transition).

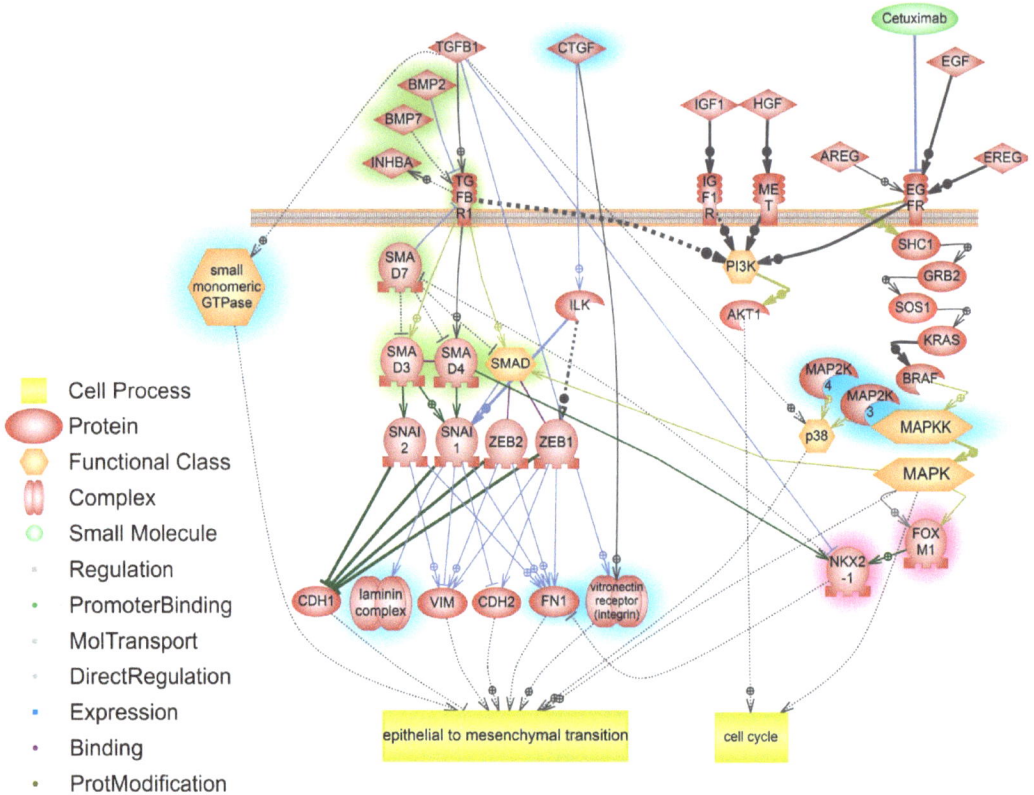

Figure 4: A detailed pathway for cetuximab sensitivity based on literature data. Entities highlighted in blue belong to the C11 cluster (epithelial-to-mesenchymal transition), highlighted in red belong to the C9 cluster (TGF-β/SMAD signaling). The right side of the pathway depicts downstream signaling from EGFR, which is inhibited by cetuximab. Cetuximab inhibits EGFR by competing with other ligands such as EGF, AREG and EREG. Activated EGFR phosphorylates SHC1, leading to the recruitment of the GRB2/SOS complex and subsequent signal propagation in cytoplasm *via* a well-characterized kinase cascade. The activity of MAPK pathway stimulates nuclear translocation of FOXM1, where the latter can activate the NKX2-1 promoter. In turn, NKX2-1 inhibits TGF-β-mediated epithelial-to-mesenchymal transition [35]. The left side of the pathway depicts the role of TGF-β/SMAD signaling in triggering the EMT process. Different ligands from the TGF-β superfamily bind with TGFBR1, which phosphorylates receptor-regulated SMAD proteins (R-SMADs) which, in turn, triggers R-SMADs binding to coSMAD - SMAD4. SMAD3 and SMAD4 cooperate to activate promoters of SNAIL1 and SNAIL2 [40]. SMADs also directly interact with ZEB1 and ZEB2. Both, SNAIL and ZEB family members repress E-Cadherin (CDH1) expression, leading to the dissolution of adherent junctions between neighboring epithelial cells. At the same time SNAIL and ZEB family members enhance the expression of mesenchymal proteins such as fibronectin and vimentin, thus activating the EMT process. Also SNAIL1 expression can be induced by integrin-linked kinase (ILK) *via* CTGF [46]. At the later stages of tumor genesis, TGF-β stimulates AKT phosphorylation *via* PI3K [47]. Consequently, tumors with activated EMT may rely on growth signals other than EGF and therefore can be unresponsive to anti-EGFR treatment.

CONCLUSION

We present a novel approach to identify putative molecular mechanisms responsible for an observed phenotype. Running the Sub-network enrichment analysis algorithm allows transformation of the large and less informative differential profiles into a smaller number of key expression regulators suggesting the mechanisms of cancer types for individual patients. Sample clustering combined with grouping of found regulators (biclustering) enables the discovery of pathways and the ability to discriminate between conditions. The approach not only enables sample classification but also provides its biological interpretation. Genes downstream of found regulators may also serve as candidate biomarkers for discriminating between various conditions.

We applied our approach to colorectal cancer to make several interesting findings. The markers of EMT and TGF-β/SMAD activated signaling are differentially expressed consistently in cetuximab non-responders. Differential expression of cetuximab-responsive *vs.* non-responsive patients identified the TGF-β pathway and the NKX2.1/FOXM1 pathway, known to inhibit TGF-β – induced events as significant regulators. Cell-line based profiling of EGFR inhibition-resistant cell lines also pointed to the TGF-β pathway and more specifically to the transcriptional machinery of EMT regulation. Our preliminary analysis of public datasets suggests that the proteins downstream from FOXM1/NKX2.1, epithelial-to-mesenchymal transition transcriptional program, EGFR and canonical/non-canonical EGFR signaling may be good candidates for biomarkers predicting anti-EGFR treatment response for colorectal cancer.

We conclude that sub-network enrichment analysis in combination with biclustering of expression regulators and samples can be used to predict the efficacy of molecularly-targeted therapies for the treatment of colorectal cancer.

MATERIALS AND METHODS

Dataset Description: We used nine microarray datasets of human CRC (GSE13294, GSE14333, GSE17538, GSE18105, GSE4107, GSE4183, GSE9348, GSE8671, GSE5851) downloaded from the NCBI GEO database [www.ncbi.nlm.nih.gov/geo]. The total number of samples equaled 863, with 784 samples from CRC patients and 79 samples from healthy donors. The GSE5851 experiment was performed using the Affymetrix HGU133A 2.0 chip while all others were performed using Affymetrix HGU133 Plus 2.0 arrays. We limited our analysis to probes that were present on both arrays. Because the HGU133A 2.0 array is a cheaper version of HGU133 Plus 2.0 array fabricated using the same technological processes, both chips have 483,674 probes where 22,585 corresponding probesets were identical.

Dataset GSE5851 contained supplementary information as to whether a patient responded to cetuximab therapy. In their work Khambata-Ford *et al.,* [41] divided patient cetuximab responses into four classes: complete response, partial response, stable disease and progressive disease. We used these additional classifications to find specific pathways and mechanisms that can explain or predict whether an individual patient will respond to cetuximab therapy.

Normalization and Probeset Filtering: Researchers admit [42,43,44] that the choice of normalization method for microarray data can significantly affect results. However, in microarray experiments that were utilized in our analysis, the authors used various approaches for normalization. For example, to make calls of expression for the GSE13294 dataset, the MAS5.0 procedure was applied while GSE18105 was normalized with robust multi-array average (RMA) method. Hence, in order to safely aggregate microarray datasets for our meta-analysis we downloaded raw expression data from all experiments in the form of CEL files and performed normalization with a single method, namely the FARMS (Factor Analysis for Robust Microarray Summarization) algorithm [45]. According to the FARMS approach, Bayesian factor analysis is done for perfect match probes, and a multivariate model is fit for each probeset. One nice feature available within FARMS framework is the ability to select only informative probesets, *i.e.,* those where most of probes of a probeset show correlated change in the concentration of mRNA across multiple arrays [48]. Therefore, FARMS can be also considered as a feature selection algorithm. Normalization and filtering of non-informative probesets by the FARMS algorithm left us with 9,254 probesets. Next, we mapped probesets to genes. In cases in which several probesets corresponded to one gene we selected only the probeset with a maximum intensity. Our final dataset consisted of 863 samples (both normal and CRC samples) and 6,377 genes.

Individual Expression Profiles: Various factors may contribute to the development of cancer: genetic predisposition, lifestyle, diet, environment, *etc.* Because of these multiple contributions, the molecular mechanisms causing one cancer differ among individual patients. Therefore, in order to discover disrupted pathways in an individual patient, it is necessary to perform SNEA for each individual sample separately. We grouped together all normal samples and estimated the cumulative distribution function F for each gene using splines [49]. For each expression value g_{ij} of i-th gene from j-th CRC sample we computed the corresponding value of the cumulative distribution function and converted it to a two-sided p-value:

$$p_{ij} = \begin{cases} 2F(g_{ij}), & F(g_{ij}) < 0.5 \\ 2(1 - F(g_{ij})), & F(g_{ij}) > 0.5 \end{cases}$$

For further analysis we used only genes with a p-value less than 0.05. Next, we calculated log-ratio between g_{ij} and mean expression of j-th gene in normal samples. To find significant regulators of genes with altered expression we ran Sub-network enrichment analysis(SNEA), in which all datasets results were limited to the top 100 subnetworks with a p-value less than 0.02.

Software. Most of the computations described in this chapter were done using the R statistical environment [www.r-project.org] and the BioConductor set of packages [www.bioconductor.org]. Sub-Network Enrichment Analysis and pathway building were performed with Pathway Studio 8.0 from Ariadne Genomics [www.ariadnegenomics.com].

CONFLICT OF INTEREST

Authors do not have any conflicts of interests with respect to chapter content.

ACKNOWLEDGEMENT

Authors thank Elena I. Schwartz for helpful discussions.

REFERENCES

[1]   Cutsem, E.V. 2007. Current treatment standards in metastatic colorectal cancer—the role of anti-epidermal growth factor antibodies. *US Oncological Disease*: 94-96.

[2]   Coutinho, A.K.; Rocha Lima, S.M. 2003. Metastatic colorectal cancer: Systemic treatment in the new millennium. *Cancer Control* 10, no. 3: 224-238.

[3]   American Cancer Society, Atlanta. 2011. Colorectal cancer facts & figures: 2011-2013. *Atlanta: American Cancer Society*.

[4]   Jemal, A., R. Siegel, J. Xu, and E. Ward. 2010. Cancer statistics, 2010. *CA: A Cancer journal for Clinicians* 60, no. 5: 277-300.

[5]   Center, M. M., A. Jemal, R. A. Smith, and E. Ward. 2009. Worldwide variations in colorectal cancer. *CA: A Cancer Journal for Clinicians* 59, no. 6: 366-78.

[6]   Abu-Shakra M, Buskila D, Ehrenfeld M, Conrad K, Shoenfeld Y. 2001. Cancer and autoimmunity: Autoimmune and rheumatic features in patients with malignancies. *Ann Rheum Dis* 60: 433-441.

[7]   Ettinger, D. S., G. Bepler, R. Bueno, A. Chang, J. Y. Chang, L. R. Chirieac, T. A. D'Amico, T. L. Demmy, S. J. Feigenberg, F. W. Grannis, Jr., T. Jahan, M. Jahanzeb, A. Kessinger, R. Komaki, M. G. Kris, C. J. Langer, Q. T. Le, R. Martins, G. A. Otterson, F. Robert, D. J. Sugarbaker, and D. E. Wood. 2006. Non-small cell lung cancer clinical practice guidelines in oncology. *J Natl Compr Canc Netw* 4, no. 6: 548-82.

[8]   Folprecht, G, Nowacki, M, Lang I, S. Cascinu, I. Shchepotin, J. Maurel, P. Rougier, D. Cunningham, A. Zubel and E. Van Cutsem. 2009. Cetuximab plus folfiri first line in patients (pts) with metastatic colorectal cancer (mcrc): A quality-of-life (qol) analysis of the crystal trial. *Journal of Clinical Oncology* Vol 27, no. No 15S: 4076.

[9]   Linardou, H., I. J. Dahabreh, D. Kanaloupiti, F. Siannis, D. Bafaloukos, P. Kosmidis, C. A. Papadimitriou, and S. Murray. 2008. Assessment of somatic k-ras mutations as a mechanism associated with resistance to egfr-targeted agents: A systematic review and meta-analysis of studies in advanced non-small-cell lung cancer and metastatic colorectal cancer. *The lancet oncology* 9, no. 10: 962-72.

[10]   Sipples, R. 2006. Common side effects of anti-EGFR therapy: Acneform rash. *Semin Oncol Nurs* 22, no. 1 Suppl 1: 28-34.

[11]   Cheng, F., S. H. Cho, and J. K. Lee. 2010. Multi-gene expression-based statistical approaches to predicting patients' clinical outcomes and responses. *Methods in Molecular Biology* 620: 471-84.

[12]   Oh, S. C., Y. Y. Park, E. S. Park, J. Y. Lim, S. M. Kim, S. B. Kim, J. Kim, S. C. Kim, I. S. Chu, J. J. Smith, R. D. Beauchamp, T. J. Yeatman, S. Kopetz, and J. S. Lee. 2011. Prognostic gene expression signature associated with two molecularly distinct subtypes of colorectal cancer. *Gut*.

[13]   Sivachenko, A. Y., A. Yuryev, N. Daraselia, and I. Mazo. 2007. Molecular networks in microarray analysis. *Journal of Bioinformatics and Computational Biology* 5, no. 2B: 429-56.

[14]   Daraselia, N., A. Yuryev, S. Egorov, I. Mazo, and I. Ispolatov. 2007. Automatic extraction of gene ontology annotation and its correlation with clusters in protein networks. *BMC Bioinformatics* 8: 243.

[15]   Novichkova, S., S. Egorov, and N. Daraselia. 2003. Medscan, a natural language processing engine for medline abstracts. *Bioinformatics* 19, no. 13: 1699-706.

[16]   de Leval, L., B. Bisig, C. Thielen, J. Boniver, and P. Gaulard. 2009. Molecular classification of t-cell lymphomas. *Critical Reviews in Oncology/Hematology* 72, no. 2: 125-43.

[17]   Lau, C. C. 2009. Molecular classification of osteosarcoma. *Cancer Treat Res* 152: 459-65.

[18] Gomez-Raposo, C., M. Mendiola, J. Barriuso, D. Hardisson, and A. Redondo. 2010. Molecular characterization of ovarian cancer by gene-expression profiling. *Gynecol Oncol* 118, no. 1: 88-92.

[19] Malhotra, G. K., X. Zhao, H. Band, and V. Band. 2010. Histological, molecular and functional subtypes of breast cancers. *Cancer biology & therapy* 10, no. 10: 955-60.

[20] Alon, U., N. Barkai, D. A. Notterman, K. Gish, S. Ybarra, D. Mack, and A. J. Levine. 1999. Broad patterns of gene expression revealed by clustering analysis of tumor and normal colon tissues probed by oligonucleotide arrays. *Proc Natl Acad Sci U S A* 96, no. 12: 6745-50.

[21] Salvador, Stan and Philip Chan. 2004. Determining the number of clusters/segments in hierarchical clustering/segmentation algorithms. Paper presented at Proceedings of the 16th IEEE International Conference on Tools with Artificial Intelligence.

[22] Miller, G. J., K. A. Bauer, D. J. Howarth, J. A. Cooper, S. E. Humphries, and R. D. Rosenberg. 2004. Increased incidence of neoplasia of the digestive tract in men with persistent activation of the coagulant pathway. *J Thromb Haemost* 2, no. 12: 2107-14.

[23] Vossen, C. Y., M. Hoffmeister, J. C. Chang-Claude, F. R. Rosendaal, and H. Brenner. 2011. Clotting factor gene polymorphisms and colorectal cancer risk. *Journal of clinical oncology: official journal of the American Society of Clinical Oncology* 29, no. 13: 1722-7.

[24] Kockar, C., O. Kockar, M. Ozturk, M. Dagli, N. Bavbek, and A. Kosar. 2005. Global fibrinolytic capacity increased exponentially in metastatic colorectal cancer. *Clin Appl Thromb Hemost* 11, no. 2: 227-30.

[25] Dugan, T. A., V. W. Yang, D. J. McQuillan, and M. Hook. 2006. Decorin modulates fibrin assembly and structure. *The Journal of biological chemistry* 281, no. 50: 38208-16.

[26] Bellam, N. and B. Pasche. 2010. TGF-beta signaling alterations and colon cancer. *Cancer treatment and research* 155: 85-103.

[27] Gulubova, M., I. Manolova, J. Ananiev, A. Julianov, Y. Yovchev, and K. Peeva. 2010. Role of TGF-beta1, its receptor TGFbetaRII, and smad proteins in the progression of colorectal cancer. *International journal of colorectal disease* 25, no. 5: 591-9.

[28] Feagins, L. A. 2010. Role of transforming growth factor-beta in inflammatory bowel disease and colitis-associated colon cancer. *Inflammatory bowel diseases* 16, no. 11: 1963-8.

[29] Muraoka-Cook, R. S., N. Dumont, and C. L. Arteaga. 2005. Dual role of transforming growth factor beta in mammary tumorigenesis and metastatic progression. *Clinical cancer research: an official journal of the American Association for Cancer Research* 11, no. 2 Pt 2: 937s-43s.

[30] Yeasmin, S., K. Nakayama, M. T. Rahman, M. Rahman, M. Ishikawa, A. Katagiri, K. Iida, N. Nakayama, and K. Miyazaki. 2011. Loss of MKK4 expression in ovarian cancer: A potential role for the epithelial to mesenchymal transition. *International journal of cancer. Journal international du cancer* 128, no. 1: 94-104.

[31] Kang, Y., H. Hebron, L. Ozbun, J. Mariano, P. Minoo, and S. B. Jakowlew. 2004. Nkx2.1 transcription factor in lung cells and a transforming growth factor-beta1 heterozygous mouse model of lung carcinogenesis. *Molecular carcinogenesis* 40, no. 4: 212-31.

[32] Li, C., N. L. Zhu, R. C. Tan, P. L. Ballard, R. Derynck, and P. Minoo. 2002. Transforming growth factor-beta inhibits pulmonary surfactant protein b gene transcription through SMAD3 interactions with NKX2.1 and HNF-3 transcription factors. *The Journal of biological chemistry* 277, no. 41: 38399-408.

[33] Xing, Y., C. Li, L. Hu, C. Tiozzo, M. Li, Y. Chai, S. Bellusci, S. Anderson, and P. Minoo. 2008. Mechanisms of tgfbeta inhibition of lung endodermal morphogenesis: The role of tbetarii, smads, NKX2.1 and PTEN. *Developmental biology* 320, no. 2: 340-50.

[34] Missero, C., M. T. Pirro, and R. Di Lauro. 2000. Multiple ras downstream pathways mediate functional repression of the homeobox gene product TTF-1. *Molecular and cellular biology* 20, no. 8: 2783-93.

[35] Saito, R. A., T. Watabe, K. Horiguchi, T. Kohyama, M. Saitoh, T. Nagase, and K. Miyazono. 2009. Thyroid transcription factor-1 inhibits transforming growth factor-beta-mediated epithelial-to-mesenchymal transition in lung adenocarcinoma cells. *Cancer research* 69, no. 7: 2783-91.

[36] Park, H. J., G. Gusarova, Z. Wang, J. R. Carr, J. Li, K. H. Kim, J. Qiu, Y. D. Park, P. R. Williamson, N. Hay, A. L. Tyner, L. F. Lau, R. H. Costa, and P. Raychaudhuri. 2011. Deregulation of foxm1b leads to tumour metastasis. *EMBO molecular medicine* 3, no. 1: 21-34.

[37] Coldren, C. D., B. A. Helfrich, S. E. Witta, M. Sugita, R. Lapadat, C. Zeng, A. Baron, W. A. Franklin, F. R. Hirsch, M. W. Geraci, and P. A. Bunn, Jr. 2006. Baseline gene expression predicts sensitivity to gefitinib in non-small cell lung cancer cell lines. *Molecular cancer research: MCR* 4, no. 8: 521-8.

[38] Albitar, L., G. Pickett, M. Morgan, J. A. Wilken, N. J. Maihle, and K. K. Leslie. 2010. EGFR isoforms and gene regulation in human endometrial cancer cells. *Molecular cancer* 9: 166.

[39] Oliveras-Ferraros, C., A. Vazquez-Martin, S. Cufi, B. Queralt, L. Baez, R. Guardeno, X. Hernandez-Yague, B. Martin-Castillo, J. Brunet, and J. A. Menendez. 2011. Stem cell property epithelial-to-mesenchymal transition is a core transcriptional network for predicting cetuximab (erbitux) efficacy in kras wild-type tumor cells. *Journal of cellular biochemistry* 112, no. 1: 10-29.

[40] Herranz, N., D. Pasini, V. M. Diaz, C. Franci, A. Gutierrez, N. Dave, M. Escriva, I. Hernandez-Munoz, L. Di Croce, K. Helin, A. Garcia de Herreros, and S. Peiro. 2008. Polycomb complex 2 is required for e-cadherin repression by the snail1 transcription factor. *Molecular and cellular biology* 28, no. 15: 4772-81.

[41] Khambata-Ford, S., C. R. Garrett, N. J. Meropol, M. Basik, C. T. Harbison, S. Wu, T. W. Wong, X. Huang, C. H. Takimoto, A. K. Godwin, B. R. Tan, S. S. Krishnamurthi, H. A. Burris, 3rd, E. A. Poplin, M. Hidalgo, J. Baselga, E. A. Clark, and D. J. Mauro. 2007. Expression of epiregulin and amphiregulin and k-ras mutation status predict disease control in metastatic colorectal cancer patients treated with cetuximab. *Journal of clinical oncology: official journal of the American Society of Clinical Oncology* 25, no. 22: 3230-7.

[42] Fujita, A., J. R. Sato, O. Rodrigues Lde, C. E. Ferreira, and M. C. Sogayar. 2006. Evaluating different methods of microarray data normalization. *BMC bioinformatics* 7: 469.

[43] Kreil, D. P. and R. R. Russell. 2005. There is no silver bullet - a guide to low-level data transforms and normalisation methods for microarray data. *Brief Bioinform* 6, no. 1: 86-97.

[44] Steinhoff, C. and M. Vingron. 2006. Normalization and quantification of differential expression in gene expression microarrays. *Brief Bioinform* 7, no. 2: 166-77.

[45] Hochreiter, S., D. A. Clevert, and K. Obermayer. 2006. A new summarization method for affymetrix probe level data. *Bioinformatics* 22, no. 8: 943-9.

[46] Larue, L. and A. Bellacosa. 2005. Epithelial-mesenchymal transition in development and cancer: Role of phosphatidylinositol 3' kinase/AKTpathways. *Oncogene* 24, no. 50: 7443-54.

[47]   Liu, X. C., B. C. Liu, X. L. Zhang, M. X. Li, and J. D. Zhang. 2007. Role of ERK1/2 and PI3-k in the regulation of CTGF-induced ilk expression in HK-2 cells. *Clin Chim Acta* 382, no. 1-2: 89-94.

[48]   Talloen, W., D. A. Clevert, S. Hochreiter, D. Amaratunga, L. Bijnens, S. Kass, and H. W. Gohlmann. 2007. I/ni-calls for the exclusion of non-informative genes: A highly effective filtering tool for microarray data. *Bioinformatics* 23, no. 21: 2897-902.

[49]   Stone, JC; Hansen, M; C. K and Y.K. T 1997. The use of polynomial splines and their tensor products in extended linear modeling (with discussion). *Annals of Statistics* 25: 1371-1470.

Index

A

ABC cassette glycoproteins 88
ABCA1 protein 97
ABCB4 protein 98
ABCB11 95, 99-101
ABCC1 protein 98
ABCC2 protein 97
ABCC3 protein 97
ABCC4 protein 98
ABCG2 protein 97
ADCYAP1 39-41
adhesion 58-9
ADIPOQ protein adiponectin 146
ADM 40-1
ADORA2A 39-42
ADRBK1 40-1, 162
AGER 39-40, 162
AGT 39-41, 120, 162
AGTR2 41, 162
AH70 cells 58
AIFM1 162
Akt 7, 14-15, 37, 109, 123, 141
AKT1 39-40
Alagille syndrome 101-2
ALOX5 protein arachidonate 5-lipoxygenase 146
ALOX15 protein arachidonate 15-lipoxygenase 146
ALS (Amyotrophic lateral sclerosis) 24-9, 31, 33, 35-7, 44
AMBP 39-41
AMPA 28, 33, 43, 55
AMPK 113, 115, 120-1
amyotrophic 24, 28, 31, 35-7
androgen deprivation therapy 3-6, 9, 14, 17-19
androgen receptor (AR) 3, 6-17, 19, 21-2
androgen stimulation 13-14
angiogenesis 60-1, 70
        positive regulation of 61, 70
annexin II 131, 135-7, 141, 143
anti-EGFR therapy 151-3, 169
antibodies, monoclonal 58, 152

antigen 3, 10, 18, 58-60, 147
apoptosis 3, 7, 10-17, 20, 22, 56-7, 60-1, 64, 66-8, 71, 101, 126
apoptosis, positive regulation of 64, 66
AQP4 39, 162
AR *see* androgen receptor
ARF6 39, 41-2
ARHGD 44
Ariadne Genomics i, iii, vi-3, 7, 24, 73-4, 104, 121, 131, 151, 168
Arthrogryposis 99-100
ATP 27, 30, 33, 39, 42, 45, 70, 95, 100-1, 163
ATPase activity, positive regulation 69, 71
ATP binding 45, 49, 54, 56, 63, 67-8, 71-2
ATPase 27, 31, 33, 45, 69, 71

B

BA import 97-8
BAX 41, 162-3
BCL2 39, 41-2
BDNF (Brain-derived neurotrophic factor) 31, 33, 37, 39-41, 46
behavioral response 57-8
bile acid (BA) 73, 76-95, 97, 99, 101-3
bile acid circulation *see* bile acid
bile acid metabolism *see* bile acid
bile acid synthesis 76-86, 89-90, 101, 103
Biochcmistry 6, 95
biological pathways v, 154
biomarker specificity 86-7, 98
biomarker test 152
biomarkers iii, vi, 73, 84, 86, 90, 104, 131, 137-8, 151, 154, 166
BMP2 40, 158-9
BMP7 158-9, 163-4
BSG 39-41
BTK 40-1

C

C-protein 110
Ca 27, 33, 42, 44-50, 55, 57, 61-2, 64, 67, 69-72, 169
receptor-mediated 46, 70
Ca-ATPase 27
calcium 27-8, 44, 55, 70, 104, 124
calcium-binding proteins 27, 36

calcium ion transport 45-6, 61
calcium mobilization 32, 44-5, 47-8, 50-3, 58-61, 63, 65, 67-8, 72
  intracellular 50, 61, 68, 70
  kinase-mediated 52
  receptor-mediated 51, 72
calcium permeability 28
calcium release 48, 57-8
cancer v, 4, 7, 9, 18-19, 21, 74, 89, 131, 151-2, 154-5, 168-71
cancer, colorectal 151-2, 155, 158, 161, 166, 169-70
cancer development   131-2
candidate biomarkers 90, 137-8, 166
canonical pathways vi, 73, 86, 116, 134-5, 140-1
CAR 75-6
CASP3 39-42, 164
CASP8 39-41
CASP9 39-40
CAT 40-1
CCL 29, 39-41, 47, 147
CCL2 39-40, 159
CCL4 40-1
CCL5 29, 40-1, 159
CCND1 39, 163-4
CCR1 29, 40-1
CCR3 40-1
CCR4 41-2, 162
CCR7 39, 41
CD1D 39-42
CD2 39-41
CD3E 41
CD4 39-41, 113, 129
CD5 41
CD8A 39-41
CD9 29, 41
CD14 39-42, 159
CD18 58
CD19 40-1
CD22 41-2
CD27 41
CD36 39-40, 42
CD38 39-41, 140
CD40 39-42
CD40LG 40-1, 162
CD44 39-41

CD45 52
CD47 41
CD48 39-42
CD53 41
CD63 40-1
CD72 29, 41, 48
CD73 66
CD79A 41
CD80 39, 41
CD81 40-1
CD86 39-41
CD99 39
CDH1 39-40, 159, 163-5
CDH2 163-4
CDKN1A 39-42
CDKN1B 41-2
CDKN2C 41
CEBPB 40-2, 121, 159, 164
cell adhesion 44-5, 49-50, 58-61, 68, 70
cell-cell adhesion 58-61
cell cycle 3, 7, 14, 19, 61
cell differentiation 53, 64
cell motility 15
cell proliferation, positive regulation of 51, 65
cell surface receptor 49-50, 53, 60
cellular processes ii, 35, 105
central nervous system (CNS) 28, 37, 46, 72
cetuximab 151-2, 158, 161, 163, 165, 169, 171
CFD 39-40
CFLAR 39, 41-2
CGP 65
chemokine activity 48, 50
cholestasis iii, 73-6, 78, 82-5, 87-91, 93-4, 99-102
cholestasis model 76, 78-80, 82, 84-8, 91-2, 97
cholestasis syndrome 99-100
Chronic Helicobacter infection *see* Helicobacter infection
CIITA protein class II 147
CK1alpha 49
CLDN3 163-4
CMA1 40
CNR2 39-41
colorectal cancer 151-2, 155, 158, 161, 166, 169-70
CR1 39

CR2 39-41
CRC 151-2
CRP 39-41
CSF1 39-40, 42, 120
CSF1R 40
CSF2 39-42
CSF3 39-41
CTGF 39, 113, 159, 162, 164-5
CTNNB 39-41
cTnT 71
CTSG 39-41
CTSS 39-42
CX 29, 40-1
CXCL1 protein chemokine 147
CXCL13 protein chemokine 147
CXCR4 29, 40-1, 50, 163
CYCS 39
CYP 41, 73, 78-80, 82-4, 86-90, 92-4, 96-8, 101, 162
cytochrome 73, 92
cytokine activity 47-8, 50
cytokines 8, 50, 85, 112, 119-20
Cytoplasm 29, 135, 146-8, 149-50, 165
cytoplasmic granule 146, 148
cytoplasmic proteins 108, 133
cytoskeleton 71, 108, 149-50

D

data analysis 104-5
database iii, 25, 35, 74-5, 77-9, 81-4, 87-9, 91, 106, 133-4, 137, 140, 146
DGC (dystrophin-glycan complex) 107-9, 111
disease
molecular mechanisms of 153
progressive 160, 162, 167
        models vi, 104-6
        pathogenesis 104
DLL1 41-2
DMD (Duchenne muscular dystrophy) 104-7, 109-13, 115-17, 119-23, 125-30
drug development 73, 89
drug-induced cholestasis 73-5, 83-4, 87, 89-91, 93, 99
drug responders 153, 162
drug targets 91, 104, 112, 115, 131
drug therapy 73, 84, 89-90, 105

E

EGF (epidermal growth factor) 8, 39, 41, 163, 165, 168
EGFR 39-41, 68, 151-2, 162-6
EGR1 39-40
ELA2 39-41
ELN protein elastin 148
EMT 151, 161, 163-6
epithelial 53, 159, 161, 170
epithelial cells 15-16, 134, 142, 161
gastrointestinal 134, 142
secretory 10-11
EPO 40-1
ER (endoplasmic reticulum) 16, 27, 29, 42, 67
ESR1 40-1, 100, 162
ESR2 39-40
ESRRA 104, 120
ETV1 protein 148
expression profile 138
differential 154-5, 162
expression regulators vi, 116, 119-21, 132-3, 146, 151, 155-6, 166
expression target pathway 116-17, 121
expression targets 116, 137, 139

F

FADD 39-40, 42
fALS 24-31, 34, 39
FARMS 167
FAS 39-42
FASLG 39-41
FCER1A 40-1
FCER1G 29, 40-1
FCER2 41
FCGR1A 29, 39-41
FCGR2 41, 52
FCGR2A 39-40
FCGR2B 29, 39-41
FK506 binding proteins 12, 21
FKBP51 13
FLOT1 40-1
FLT3LG protein 148
FOXA 163

FOXH1 120-1
FOXJ1 41
FOXM1 162-3, 165
FOXP3 162
FPR1 39-41
FPR2 40-1
fractalkine 50

G

G-protein 46, 48-51, 56-8, 65, 67, 69-70, 111
regulation of 56, 69-70
GAB2 40-1
gamma-aminobutyric acid secretion 57
GAS6 40
GAST 40-1, 159
gastric cancer 131-2, 135-8, 142-3
Gastric Cancer Model iii, 133, 135, 137, 139, 141, 143, 147, 149
gastric carcinoma 143-4
gastrin 9, 131-7, 140-2, 146
gastrin receptors 135, 142
Gastroenterology 141-2
GDI (GDP dissociation inhibitor) 33, 44-5
gene expression v, 24, 29, 63, 94-5, 113, 116, 125, 128, 144, 151, 155-6, 170
Gene Expression Omnibus (GEO) 25, 132
gene sets 154
GEO database 34
GHRL 40
glomerular lesions 102
glomerular mesangiolipidosis 102
glucoronidate 91
GluR2 28, 55
glutamate 28
glutamate receptors 24, 55
Glycoprotein 122
glycoprotein binding 49, 60-1
GNAI2 41-2
GNAQ 40-1
GPR44 40-1
GRAP2 40-1
GRIA2 30-1, 43
GRIA2 glutamate receptor 33, 55
GRIN1 28, 30, 40, 42

GRIN1 glutamate receptor 33, 55
GRK6 32, 34, 41, 56
GRM2 protein glutamate receptor 148
GSE4342 163
GSE4390 34
GSE5851 155, 158, 162, 166
GSE16362 34
GSE18105 166-7
GSE20854 163
GSEA (gene set enrichment analysis) 154
GTP binding 68-9
GZMB 29, 39-41

H

HBP 61
HCK 40-1
Helicobacter 131-5, 137, 142
Helicobacter-associated gastric corpus carcinogenesis 142
Helicobacter Felis Infection 132
Helicobacter infection 131-8, 142
Helicobacter pylori 142
heparin binding 47, 53-4, 60, 70
hepatic 75-6
hepatocytes 73, 76-80, 82-3, 88, 90
hereditary cholestasis 74-5, 84, 87, 95, 99
HGF 39-41, 159, 163
HGF/SF 15
high-throughput i-ii
HMOX1 39-41, 163
HNF1A 40, 75-6, 159
HNF4A 75, 80, 82, 90
homologs 79, 147
hormone activity 59, 63, 65
hormone secretion 65
host cell surface receptor 58-9
HSD 101
Hsp90-binding protein 21
HSP90B1 protein heat shock protein 148
HSPD1 41
HSPD1 protein spastic paraplegia 148
hTERT 61
Human gastric carcinogenesis 142

hydrolase activity 45, 56, 63-4, 66-7
hypergastrenimia 131, 138
hypcrsensitivity 51-3

I

ICAM1 29, 39-41, 159
ICAM1-dependent adhesive interaction 58
IGF-1 8, 11, 31, 119
IGF1 31, 34, 39-41, 59, 120-1
IGF2 39-40
IL 40-1, 159, 162
IL-6 8
IL1A 40-2, 159
IL1B 39-42, 87, 138
IL2 39-42, 138
IL2RA 41
IL4 39-42, 120
IL4R 40
IL5 40-1
IL6 39-42, 117, 119, 138
IL8 39-40, 163-4
IL9 40-1, 159
IL10 39-42, 120, 138
IL10RA protein interleukin 148
IL11 40, 159, 162
IL11 protein interleukin 148
IL13 40-1, 159, 162
IL15 39-42, 162
IL16 41
IL17A 39-42, 162
IL17A protein interleukin 148
IL18 40-1
IL18 protein interleukin 148
immune response 29, 44, 47, 49-50, 52, 68, 72, 159
immunophilin protein family 13
inflammation 25, 29, 93, 106-7, 112, 115, 129
Inflammatory bowel diseases 170
inflammatory response 25, 48-51, 60-1, 63, 65-8
INHBA 39, 41-2, 158-9, 163-4
inhibition of CD18 58
inhibition of HNF4A 90
INPP4B 14

INPP5D 40-1
INS 39, 41-2, 120
INS-GAS mice 132, 142
interleukin-6 8, 143
ion channel activity 55, 62
ion transport 45, 55, 61-2
iressa 163
ITGA 39-40, 163
ITGAL 29, 41
ITGAM 29, 39-41
ITGB1 39-41
ITGB2 29, 39-41
ITGB2 integrin 60-1
ITPR 27, 29-30, 34, 36, 39, 41-2, 61

J

JAG1 100-2, 163
JAK2 40-1
JAK3 40-1
JAMA 18-19

K

kallikreins 10-11
kinase 22, 124, 135, 141-2, 163
kinase activity 49, 54, 56, 62-4, 66-8, 72, 123
KIT 40-1
KITLG 40-1
KLF4 39, 41-2
KLK3 10
KLRA1 41
KLRK1 29, 39-41
KNG1 90, 159
Knowledge Networks iii-vi, 74, 76, 132, 134, 136, 138, 140, 142, 144, 146, 148, 150
KRAS 39-40, 100, 151
Kupffer cells 58-9

L

LAMA1 162-3
LAT 40-1
LAT2 40-1

LEP 39-41, 120-1, 159
leukocyte 48, 50, 58-9, 61
LIF 39-42, 159
ligands 10, 47-8, 50, 83, 85, 93, 101, 116, 147-8, 150, 158, 165
LTB4R 40-1
LTF 39-41
LXR 120-1
Ly-GDI 44-5

M

macrophages, incubation of 61
MAP, heat 138, 151, 160-1
MC2R 162
MDR1 73, 88-9, 92
mdx mice 109-10, 112-13, 115, 124-7, 129
mdx mouse 125, 129-30
membrane 58, 98, 109, 150
membrane docking 58-9
mesenchymal cell proliferation 53-4
metabolic process 45-7, 63, 65
metabolism 74, 76, 78-9, 82, 95, 113-14
MHC (major histocompatibility complex) 13, 147
MHC class 63
MHC molecule 58-60
microarray data 26, 130, 167, 171-2
mitochondria biogenesis 104, 113, 119
MLXIPL 120-1
MME 39, 162
MMP1 163-4
MMP9 31, 39-42, 163
mouse models 25, 27, 36, 142
MS4A1 41
MSR1 40
muscle cell 107-8
muscle contraction 71, 109, 123
muscle degeneration 110, 113, 125, 127
muscle fibers 123-4, 127, 130
muscle remodeling 104, 113-15, 120, 129
muscle repair 112, 115, 126
muscle weakness 105
muscular dystrophies 105, 122, 125, 127, 129
MYD88 39-42, 143

MYLK 41
myofibrillar genes 112, 128
myogenesis 112, 119

N

N-methyl D-aspartate 33
NADPH 43
natural language processing (NLP) 35
NCAM1 39
NED 9
negative regulation 7, 14, 49, 53-4, 64-5, 67, 69, 71
negative regulation of apoptosis 67, 71
negative regulation of inflammatory response 66-7
NEU1 39, 41-2, 162
Neurogenetics 128-9
Neurological Sciences 128-9
neutrophils 113
NF-kappaB 142-3, *see also* NF-κB
NF-κB 13, 109, 112, 114, 120, 132-41
NF-κB activation 132, 134-6, 140-1
NF-κB signaling pathway 132, 134, 141
NF-κB super-activation 137-8
NFAT cell activation 67
NFKB 41, 43, 137, 139
NGF 39-41
nicotine 57-8
nitric oxide 104, 126
nitric oxide synthase *see* NOS
NMD 122, 124, 128-9
NMDA 28, 42-3
NMDA receptor activation 28, 37
NMDAR-mediated calcium influx 55-6
NOS2A 39-41
NPY 40-1
NR 42, 159, 162-3
NRF1 120-1
NT5E 29, 41, 66, 162
NTRK2 39
nuclear factor-kappaB, 20-1, 127, *see also* NF-κB and NF-kappaB
nuclear hormone receptors 95, 159
nuclear receptors 76, 83, 89
nucleotide binding 45, 49, 54, 56, 62, 64, 66-70, 72, 88-9, 95

O

oncogene 3, 21, 94, 143, 171
organic anions 91
OSTalpha protein 98
OSTBETA protein 98
ovarian cancer 170
over-expression 8, 13, 20, 30, 50, 83, 90, 110, 112, 114-15, 125

P

P-glycoprotein 95-6, 101-2
PARP1 39-40
pathogenesis 106, 111, 113, 161
pathogenetic mechanisms, known 101-2
pathology 93, 113, 124, 127-9
pathway activity 151, 163
pathway analysis 24, 35, 73, 131, 151
pathway building 105, 168
Pathway Studio i-vi, 3, 7, 24, 32, 34-5, 38, 73, 77, 84-8, 117, 119, 131, 139, 151
patient response 151, 160
patient stratification 5, 151, 153
PAX2 162
PBSF/SDF-1 50
PDCD1 40-1
PDGFA 162
PDGFR (platelet-derived growth factor receptor) 68
PECAM1 40-1
peptidase activity 57, 66-7
serine-type 57, 66-7
PF4 39-42
PGF 40-1
phagocytosis 37, 51-3
Pharmacophore models 92
phosphatidylinositol 14, 171
phosphatidylserine 44
phospholipase 51-2, 57-8, 70, 136
activation of 47, 57-8
phosphoprotein binding 49
phosphorylation 8, 11, 13, 21, 49, 56, 72, 123, 143
PI3K 123, 137, 141, 165
pigmentation 62
PIK3CD 40-1

PINA database 140-1
PKC-delta 14, 137, 140
PKC-delta and TRAF6 proteins 140
Plasma membrane 33, 45, 69, 123, 135-6, 141, 146-8, 149-50
plasma membrane receptors 133
PLAT 39-41
platelet-activating factor receptor 31, 34, 67
platelet-derived growth factor receptor (PDGFR) 68
PLAU 39-41, 138, 163
PLAUR 39-41
PLC 57, 72
PLCG1 39-41
PLCG2 40-1
PLD1 40
PLD2 40-1
PLG 39-41
PMCA2 45-6
PMCA2b 45
polymorphisms 99-100, 102
POMC 40-1
positive regulation of mesenchymal cell proliferation 53-4
positive regulation of type III hypersensitivity 52
postsynaptic density protein PSD-95 123
POU 41, 149, 159
PPARA 39-42, 75-6, 80, 120
PPARG 39-42, 120
PPARGC1A 113, 115, 120
PPP2CB protein protein phosphatase 149
PRDM1 41
Predicting Cetuximab Sensitivity 153, 155, 157, 159, 161, 163, 165, 167, 169, 171
PRKCA 39-41, 164
PRKCB 39-41
PRKCD 39-43, 137
PRKCD protein kinase 66
PRKCQ 41, 43
PRKCQ protein kinase 67
PRL 41, 159
PRNP 39-41
PROC protein 67
prostate 3, 6, 8-9, 13, 16, 18, 20-1
prostate cancer 3-9, 15-22, 152
advanced 3-6, 17
　　　cells 6, 8-9, 13-14, 20-2

metastatic 4-5
overview pathway 3, 7
progression 8-9, 11, 16, 20, 22
treatment 3, 5, 18-19
prostate tissue 4
prostatectomy, radical 4-5, 18
protein beta-2-microglobulin 146
protein binding 30, 44-50, 52-5, 57-64, 66-72, 149
calcium-dependent 70
protein-breakdown, accelerated 112
protein
beta-2-microglobulin 13
calcineurin 59
    chemokine 147
druggable 136
excitotoxicity-related 31
extracellular 137
heterotrimeric 69
over-expression of 138
regulatory 69
protein components 76
protein connectivity 80
protein-coupled receptor 13, 126
protein entities 76, 85
protein homology 73
protein homooligomerization 70
protein import 71
protein kinase 12-15, 21-2, 51, 64, 67, 72, 109, 111, 124, 142
mitogen-activated 64
protein kinase activity 49, 54, 56, 62, 64, 66-8, 72
protein kinase binding 54-5, 61, 68
protein kinase C (PKC) 12, 14-15, 21-2, 51, 64, 67, 72, 109, 124, 140, 142-3
protein localization 55
protein loss 13
protein networks 169
protein oligomerization 62
protein phosphorylation 50
protein product 122
protein-protein interaction networks 94
protein-protein interactions 142
protein targets 73, 76, 82, 84-5
protein transport 68
protein tyrosine phosphatase activity 37

protein tyrosine phosphorylation 52
PRTN3 39, 41-2
PSA (Prostate-specific antigen) 5, 10
PTAFR 31, 34, 40-1, 67
PTEN 14, 39-41, 171
PTGER1 40-1
PTGER1 protein prostaglandin 149
PTGER2 40-1
PTGER2 protein prostaglandin 149
PTGS2 39-41, 164
PTH 39-41
PTK2B 39-40, 42
PTK2B PTK2B protein tyrosine kinase 68
PTMA protein prothymosin 149
PTPN6 40-1
PTPN11 39-41
PTPRC 39-42
PTPRJ 40-1
PXR 75-8, 80, 83, 89-90, 92-3
nuclear receptor 94-5
PXR activation 90-2, 95
PXR-FXR proteins 83

R

r-ras 32, 34, 69
RAC1 39-40
radiation therapy 4-5, 18
radiotherapy 5
RAF1 39-40
reactive oxygen species (ROS) 112
receptor activation 96
receptor activity 47-58, 60-3, 65, 67-8, 70
chemokine 48, 50
receptor binding 46, 49, 51, 65
receptor expression, activating factor 37
receptor pathway 137
regulation of cell 44, 61
regulation of leukocyte 58
regulation of neuron apoptosis 46, 56
regulation of synaptic plasticity 47
REL 39-42, 137
REL protein v-rel reticuloendotheliosis 149

RELA 40-1, 43

ResNet 29, 35, 39-40, 74, 77-8, 81-2, 84, 90, 92, 106, 133, 137, 140

RGS7 32, 38, 41, 69

RGS7 regulator of G-protein signaling 34, 69

Rho GDP dissociation inhibitor 44

Rho GDP-dissociation inhibitor activity 44

Rho GTPases 44-5

potential regulator of 44-5

Rho protein signal transduction 44-5

RhoA 39-41, 120, 135-7

RhoA proteins 140

ROCK1 162

RUNX 40, 86, 162, 164

RXR 75-6, 90

RYR1 41

S

SAC (Stretch-activated channels) 109, 124

sarcolemma 108-10, 122-3

sarcoplasmic reticulum (SR) 27

scaffolding 111

secreted proteins 73, 77, 85, 90, 133, 138, 149

secretion 77

hormone 65

SELP 39, 41-2

SELP protein selectin 149

SEMA4D 41-2

semantic triplets iv

semenogelins 11

sensitivity profiles 163-4

serine/threonine protein kinases 14

serine-type endopeptidase activity 57, 66-7

serotonin 57

serotonin 5-HT 57

serotonin binding 57

SERPINA1 41, 100-1

SERPINA4 protein serpin peptidase inhibitor 149

SERPINE1 39, 163-4

SERPING1 protein serpin peptidase inhibitor 149

serum 14, 22

Serum/glucocorticoid-induced protein kinase-1 22

SFK (Src family kinase) 62

SFTPB 162
SFTPC 162
SFTPD 40-1, 162
SGK1 14
SH 40-1
SHP (small heterodimer partner) 75-6, 80, 82, 86, 90
signal transducer activity 47-51, 56-7, 63, 65, 68-9
signal transduction 19, 47, 49-53, 55-7, 60-1, 64-6, 68-70, 72, 111, 126, 133
signal transduction proteins 86
signaling cascades 6, 8-9, 17, 67, 105
signaling pathways 47-8, 50, 53, 57-8, 60-1, 63-4, 72, 80, 86, 116, 123, 129, 160
canonical 116, 140
signaling protein activity 51
skeletal muscle 28, 105-6, 119, 123-30
SLC6A2 162
SLC8A1 39-41
SLC10A1 protein 98
SLC22A1 protein 98
SLC22A8 protein 97
Slco1a1 protein 98
SLCO1A2 protein 98
SLCO1B1/OATP2 protein 97
SLCO1B3/OATP4 protein 97
SLCO4C1 protein 98
SMAD 8, 20, 39, 41, 158-9, 164-5, 170
SMAD-interacting protein 161
SNAIL 164-5
SNCA protein synuclein 149
SNEA (sub-network enrichment analysis) ii, v-vi, 25, 29, 33, 35, 39, 132-3, 137, 139,
        146, 151, 153-9, 163, 165-9
SNEA algorithm 155
SNEA analysis 139, 146, 158
SOC (Store-operated channels) 109
SOCS 40-1, 149
SOD1 24-5, 35, 40
SOD1-G93A mice 25, 29
SOD1-mutant mice 30
SPARC protein 149
SPHK1 40-1, 46
SPHK2 40
SPI1 39-42, 120
SPN 41
SPN protein sialophorin 149

SPP1 39-41
SR (sarcoplasmic reticulum) 27
SRC 39-41, 68
Src family kinase (SFK) 62
SREBF 80, 104, 120-1
SST 41-2
ST 40-1
STAR 162
STAT 8, 39-42
Store-operated channels (SOC) 109
Stretch-activated channels (SAC) 109, 124
Structure of P-glycoprotein 96
sub-networks 29, 32, 85, 104, 119, 139, 154
subnetworks, *see* sub-networks
symbiont 58-9
synaptic plasticity 47, 56, 64
synaptic transmission 55-7, 65, 69-70
synergistic action 133, 135, 137, 139-40
Synergistic Carcinogenesis 131
synergy 131, 134, 146

T

T-cell activation 143
TAB1 protein 141
TAC1 40-2, 159
TACR1 40-2
target genes 38, 156, 158, 163
regulating 13
target tissue 88
TBXA2R 41
TGF-b and EGFR pathways 164
TGF-b/SMAD pathway 151, 159, 161
TGF-b/SMAD signaling 164-5
TGF-beta 8, 20, 170
TGFA 39-40
TGFB 39-42, 98, 113, 118, 120, 159-60, 162-3
TGFBR1 40, 158-9, 165
THBS1 162, 164
therapy, hormonal 4, 6, 9
thrombin 51
TIRAP 140
TIRAP protein 140

tissues, normal 29, 153, 155
TLR *see* Toll-like receptor
TLR2 protein 149
TLR4 39-40, 135, 142, 159
TMSB4X protein thymosin 150
TNF 39-42, 87, 118, 120, 138
TNFRSF1A 39-40, 42
TNFRSF1B 40-1
TNFRSF4 41
TNFRSF8 40-1
TNFRSF9 protein tumor necrosis factor receptor superfamily 150
TNFRSF11B 39-40, 162
TNFRSF13C protein tumor necrosis factor receptor superfamily 150
TNFRSF21 41-2
TNFSF4 39-41
TNFSF10 39-41
TNFSF11 41-2
TNFSF18 protein tumor necrosis factor 150
TORC2 120-1
TP53 39-41, 120, 163-4
TP53 tumor protein 71
TRAF2 41-2
TRAF6 39-40, 131, 135-6, 140-3
TRAF6 activation 140
training dataset 155-6, 158
transcription 13, 20, 38, 56, 80, 159
transcription factors v, 11, 14, 24, 64, 71, 76, 80, 85-6, 92, 116-17, 133, 137, 158, 171
transcriptional repressor 7, 11, 32-3
transferase activity 49, 54, 56, 64, 67-8, 72
transgenic mice 25, 32, 34, 132
transition, mesenchymal 53, 159, 161, 170
transporters 79, 85-6, 91-3
transporting 33, 45
TRAP (thrombin receptor activating peptides) 51, 61
TRH 40-1
troponin 27, 29, 71, 110
TRPC 109, 124, 126
TRPV 40-1, 109, 125
tumor suppressor 3, 11, 13-14, 16
tumors, primary 4
type III hypersensitivity 52
TYROBP 40-2

V

Vav1, amino-terminus region of 44
VCAM1 40-1
VDR (vitamin D receptor) 39, 42, 89, 95, 163
VEGFA 39-42
VIP 39-42
virion attachment 58-9
vitamin D receptor *see* VDR
VPS33B 99
VTN 39-40
VWF 40-1, 51

W

wild-type (WT) 56, 153

X

xenobiotics 73, 90, 93-5

Z

ZAP 29, 41, 72
Zap-70 72
ZEB1 159, 163-5
zinc ion binding 65-7
ZIP1 and ZIP2 proteins 11
Zn superoxide dismutase 24-5, 35
Zn2 11

www.ingramcontent.com/pod-product-compliance
Lightning Source LLC
Chambersburg PA
CBHW041700210326
41598CB00007B/476